TIME

PAST, PRESENT, FUTURE

Nijhoff International Philosophy Series

VOLUME 30

The titles published in this series are listed at the end of this volume.

TIME
Past, Present, Future

ZDZISŁAW AUGUSTYNEK
University of Warsaw, Poland

Translated from the Polish
by Stanisław Semczuk and Witold Strawiński

SPRINGER SCIENCE+BUSINESS MEDIA, B.V.

Library of Congress Cataloging-in-Publication Data

Augustynek, Zdzisław.
[Przeszłość, teraźniejszość, przyszłość. English]
Time past, present and future / Zdzisław Augustynek.
p. cm. — (Nijhoff international philosophy series; v. 30)
Translation of: Przeszłość, teraźniejszość, przyszłość.
Bibliography: p.
ISBN 978-94-010-5648-9 ISBN 978-94-011-3706-5 (eBook)
DOI 10.1007/978-94-011-3706-5
1. Time. 2. Relativism. I. Title. II. Series.
BD638.A8313 1989
115—dc20 89-8193

ISBN 978-94-010-5648-9

Sold and distributed in Albania, Bulgaria, Chinese People's Republic, Cuba, Czechoslovakia, Hungary, Korean People's Republic, Mongolia, Poland, Romania, the U.S.S.R., Vietnam and Yugoslavia:
by ARS POLONA,
Krakowskie Przedmieście 7, 00-068, Warszawa, Poland

Translation from the Polish original
Przeszłość, teraźniejszość, przyszłość

To the memory of
Henryk Mehlberg

CONTENTS

PREFACE

This book is dedicated to the memory of Professor Henryk Mehlberg, primarily because I want to recall his name to readers in the West; for, although Professor Mehlberg was the foremost Polish philosopher in the field of philosophy of time—in that version which is related to twentieth-century physics—his fundamental work concerning time was published in English 43 years after its original publication (cf. Mehlberg, 1980). I am deeply indebted to Henryk Mehlberg. For two years after the Second World War, I was his student at Wrocław University (where he lectured on philosophy and mathematical logic). At that time, I developed, under his influence, a long-life fascination with the philosophy of time, and was taught by him how to approach these problems effectively.

My research in this field has resulted in three books. The first, *Properties of Time*, published (in Polish) in 1970, concerns the topological properties and properties of symmetry (homogeneity and anisotropy) of time. The second, *The Nature of Time*, published (in Polish) in 1975, looks at possible definitions of time within both theories of relativity. The present book is the third, originally published (in Polish) in 1979; it concerns philosophical problems of past, present, and future and is here presented in English to reach a wider audience.

My reflections about these problems were for many years the subject of my anxiety because I did not believe that they could be expressed, let alone solved, within the conceptual apparatus of relativistic physics and set theory, i.e., within apparatus in which I discussed the nature and properties of time. It seems to me now that I have finally succeeded in overcoming this difficulty, as it be manifest, I hope, to the reader of this book.

The book presents, within the framework of the Special Theory of Relativity, a fully developed and partially formalized relational and objectivistic conception of past, present, and future. Two variants of this conception are presented here: the standard version in which relative time relations are applied, and the non-standard version based on absolute time relations. Finally, there is a critical analysis of two other conceptions, one non-relational and the other subjectivistic, which stand in contrast to those developed here.

The book also contains an analysis of the relationship between past, present, future, and time itself; an attempt to clarify the possible connections between past, present, future, and existence; a time-relational characterization of the becoming of things and events; and, finally, an analogous characterization of the coming-to-be, duration, and passing away of things.

The conception introduced here assumes the ontology of what is referred to as point-eventism, according to which every object is either a point-event (spatio-temporally non-extended), or a set founded in such events. Things, for instance, are special sets of such events. The latest results obtained in this ontology are contained in my article on point-eventism (Augustynek, 1987) which is included as an appendix to this volume.

The book is addressed to students of the philosophy of science, especially the philosophy of physics, and to researchers in this field. Only an elementary knowledge of the Special Theory of Relativity and mathematical logic is required to follow it.

Lastly I would like to thank my young colleagues Dr Witold Strawiński and Stanisław Semczuk who took upon themselves the task of translation the book, and helped to improve its form. Thanks are also due to Dr Elżbieta Paczkowska, the editor of *Reports on Philosophy*, for permission to reprint here the article 'Point-Eventism'. I also thank Zofia Osek, the PWN editor, who put much effort into the editing of this book.

May 1987 *Zdzisław Augustynek*

LIST OF SYMBOLS

LOGICAL SYMBOLS

$\sim, \wedge, \vee, \rightarrow, \equiv$	negation, conjunction, disjunction, implication, equivalence
$\bigwedge, \bigvee$	universal quantifier, existential quantifier
$\in$	membership
$=$	equality
$\subset, \subsetneq$	inclusion, proper inclusion
$\cup, \cap$	union, intersection
$\bigcup, \bigcap$	union-set, intersection-set
$\emptyset$	empty set
$\{x\}$	unit-set of x
$\bar{Q}$	complement of the relation Q
$\breve{Q}$	converse of the relation Q
refl	the set of all reflexive relations
irrefl	the set of all irreflexive relations
symm	the set of all symmetrical relations
asymm	the set of all asymmetrical relations
trans	the set of all transitive relations
intrans	the set of all intransitive relations

SPECIAL SYMBOLS

S	the set of all events
$x, y, z, \ldots$	events
T	the set of all things
T′	the set of all thing-like objects
$a, b, c \ldots$	things
Pr	the set of all processes

Pr'	the set of all process-like objects
Cr	the set of all cross-sections
Cr'	the set of all cross-section-like objects
Ks	the set of all coincidents
Ks'	the set of all coincident-like objects
$\mathbf{U}$	the set of all inertial reference systems
$u, u_1, u_2 \ldots$	the inertial reference systems
W_u	relation relatively earlier
$\breve{W}_u$	relation relatively later
R_u	relation of relative simultaneity
P_x^u	relative past of the event x
N_x^u	relative present of the event x
F_x^u	relative future of the event x
$P_{a\|x}^u$	relative past of the thing a
$N_{a\|x}^u$	relative present of the thing a
$F_{a\|x}^u$	relative future of the thing a
$\mathbf{S}\|R_u$	the set of all abstraction classes of the relation R_u
$m, n, \ldots$	moments
P_m^u	the past of the moment m
N_m^u	the present of the moment m
F_m^u	the future of the moment m
$\bar{R}_u$	relation of relative non-simultaneity
$\bar{N}_x^u$	relative non-present of the event x
L_u	relative co-location
$\bar{L}_u$	relative space separation
W	relation absolutely earlier
$\breve{W}$	relation absolutely later
R	relation of absolute simultaneity (quasi-simultaneity)
P_x	the absolute past of the event x
N_x	the absolute present of the event x (quasi-present)
F_x	the absolute future of the event x
H	oriented causal relation
H'	unoriented causal relation
$P_{a\|x}$	the absolute past of the thing a
$N_{a\|x}$	the absolute present of the thing a
$F_{a\|x}$	the absolute future of the thing a
$\bar{L}$	absolute time separation
L	absolute co-location (quasi-co-location)
K	space-time coincidence

K'	space-time similarity
Ec	time extension
Ep	space extension
Cn	time continuity
Cc	causal connectivity
$A(a\|b)$	the thing *a* comes-to-be with respect to the thing *b*
$Z(a\|b)$	the thing *a* passes away with respect to the thing *b*
$D(a\|b)$	the thing *a* lasts with respect to the thing *b*
$B(a\|b)$	the thing *a* becomes with respect to the thing *b*
$A'(x\|a)$	the event *x* comes-to-be with respect to the thing *a*
$Z'(x\|a)$	the event *x* passes away with respect to the thing *a*
$D'(x\|a)$	the event *x* lasts with respect to the thing *a*
$B'(x\|a)$	the event *x* becomes with respect to the thing *a*
$A''(x\|y, v)$	the event *x* comes-to-be with respect to events *y*, *v*
$Z''(x\|z, v)$	the event *x* passes away with respect to the events *z*, *v*
$D''(x\|v)$	the event *x* lasts with respect to the event *v*
$B''(x\|y, v, z)$	the event *x* becomes with respect to the events *y*, *v*, *z*

CHAPTER I

INITIAL ASSUMPTIONS

The conception of past, present, and future discussed in this work is presented in the form of a simple system comprising axioms and a number of important and interesting theorems. Several different systems of this kind will be formulated and called theories of past, present, and future.

These systems of past, present, and future are expressed in the language of set theory. This is an initial assumption which, however, concerns more than the language applied here; it also extends to the ontological commitment suggested by set theory. More precisely, it is assumed not only that objects exist as individuals, but that their universal set, its subsets, and the sets (or families) of these subsets, etc. also exist.

The second of the initial assumptions is the ontology of eventism which—in the form proposed here—is, in my view, the only ontology adequate to both theories of relativity, which include fundamental information about time and space. In view of the importance of this ontology in this context, it will be discussed here at some length; at the same time, some of my earlier views (cf. Augustynek, 1975) will be revised.

The first and main thesis of eventism is to the effect that every object is an event or a set of events, or a set of such sets, etc. Consequently, events serve here as the only individuals or non-sets.

Events are here accepted as objects non-extended in time and space (in the ordinary sense of the words), that is, as point-events. The set of all events is said to constitute our universe (universal class), and is denoted by **S**, while its elements are denoted by x, y, z, etc.; hence $x, y, z, \ldots \in \mathbf{S}$.

The second thesis of eventism states that things (bodies in a broad sense) are certain proper subsets of the set **S**, that is, certain sets of point-events. The set of all things is denoted here by **T** and its elements by $a, b, c, \ldots$, etc.; hence $a, b, c, \ldots \in \mathbf{T}$ and, according to the above theorem, $a, b, c, \ldots \notin \mathbf{S}$.

The concept of a 'thing' is not defined here, at least not in the form of an equivalence definition; that is, it is not specified what kind of sets of events the things are. Our understanding will be in line with the physical usage of the term. Let us note at once that such objects as reference systems (including inertial reference systems which will be mentioned later) are treated in physics as things (bodies) of a specific kind.

It should be also added that various time, space, and space-time relations are determined in the set of all events **S**, as are different physical relations, for instance, the causal relation. Some of these are not defined and are treated as primitive, while others are determined by means of them. According to set theory and the above-mentioned assumptions, these relations are certain proper subsets of the Cartesian products of the set **S**.

The form of eventism presented above varies considerably from the version I formulated previously (Augustynek, 1975). The concept of a thing is basic here. In the earlier work I suggested that the major shortcoming of eventism is the lack of an equivalence event-like definition of a thing. I also expressed doubts whether such a definition could ever be formulated.

Today, I no longer believe that eventism necessarily requires the construction of such a definition. All it requires is a much weaker assumption, namely that things are sets of events, or more strictly, that they are proper subsets of the set **S** (the second thesis). What sort of sets of events they are, can be found out indirectly from a number of theorems concerning things, their general properties, connections with events, etc. Certain theorems of this kind will be considered below.

In accordance with this, a possible axiomatic system of eventism should include the following primitive terms: the symbol **S** of the set of all events, the symbol **T** of the set of all things, and—something not so far mentioned—symbols of time, space, and space-time relations, as well as symbols of certain physical relations.

The idea of constructing the theory of eventism along these lines (i.e., without an equivalence definition of a thing) was suggested to me by an excellent book on geometry (Borsuk and Szmielew, 1972). In this book the authors treat straight lines as sets of points, so that for points the relation 'lies on a line' (coincidence) is identical with the relation 'belongs to the line'. At the same time, the set of straight lines is treated as a primitive concept of geometry on a par with the set of all points, i.e., space. By analogy, the assumption that things, i.e., elements of the set **T**, are subsets of the set **S**, i.e., certain sets of events, makes it possible to identify the relation of the occurrence of an event 'in' a thing with the relation of an event belonging to a thing. Consequently, the ontology of eventism could not be constructed only if a set-theoretical explication of the above-described relation of the occurrence of an event 'in' a thing were to be impossible.

It should be stressed that a few years after the publication of the Polish version of the present book (1979) I demonstrated that an equivalence event-like definition of a thing is possible within the framework of eventism; this essay is included here as an appendix. This considerably reinforces the theoretical position of the ontology of eventism; but the definition, as was pointed out earlier, is not a necessary precondition of this ontology.

On the grounds of the adopted ontology of eventism and set theory, reality has a 'layer' structure. The fundamental layer is the set **S** of all point-events. These, as individuals, are of logical type 0, while their set, i.e., the world of events, is of logical type 1. The second layer consists of the set—let us denote it by **S′**—of all subsets of events; these subsets of course are of type 1, while their set is of type 2. The elements of the set **S′** are obviously set-theoretical parts of the set **S**. It was shown earlier that the set of things **T** is a subset of the set **S′**, i.e., that things are set-theoretical parts of the set **S**. As is known, time, for instance (defined by abstraction), is also a subset of the set **S′**, i.e., its elements (moments) are set-theoretical parts of the set **S**. The same is true of physical space and space–time (defined by abstraction). So the sets of things, time, etc. are of logical type 2, whereas their elements (things, moments, etc.) are of type 1. Let us note that ail properties of things, moments, etc. are also subsets of the set **S′**, and are therefore of type 2 (properties are identified here with the corresponding sets). The third layer consists of the set—let us denote it by **S″**—of all sets of subsets of the set of events **S**; its elements belong to logical type 2, while the set **S″** itself is of logical type 3. It will be readily noted that properties of time, space, and space–time are all subsets of the set **S″**, and belong to type 3.

One can, of course, speak of further 'layers' of the outlined set-theoretical structure of reality. In view of that, it may be asked whether this structure ends at any definite layer or logical type. I cannot give an answer to this question, nor indeed to the meta-question concerning the epistemological status of the question itself.

Our versions of the theory of past, present, and future will be formulated within the framework of the Special Theory of Relativity (STR)—the physical theory of space–time which ignores the presence of the gravitational field, and therefore distinguishes the class of so-called inertial reference systems. This is our third initial assumption.

There are several reasons for confining ourselves to STR and not entering, among other things, into the domain of the General Theory of Relativity (GTR). Firstly, STR does not give rise to any doubts with respect to its formalism, physical contents, degree of connection with experiments, or range of application. The same cannot be said of GTR, which is still (new empirical data and new ideas notwithstanding) at the stage of development.

Secondly, in STR it is assumed, at least implicitly, that the universe of events S is metrically unlimited, and that it has no boundary points (events). This is equivalent to the statement that time (defined by abstraction or by means of relations) is unlimited, and does not have any boundary points (moments); it is thus an object isomorphic with the space E^1, i.e., the Euclidean straight line. The same is assumed in STR with respect to the three dimensions of physical space. Thus the world of events S is unlimited and does not have boundary points in any of its space dimensions. Modern cosmology based on GTR presents data which favour a metrically limited model of the physical world with a singular boundary point (the beginning) in time. Because this world model has not yet been fully verified, I work on the basis of the cited premises of STR since they considerably simplify the versions of the theory of past, present, and future formulated below. Pragmatic reasons also play a role.

Thirdly, only within STR is it possible to construct the different versions of the theory of past, present, and future with which we are concerned here, and also to establish strictly definite connections between them, as well as investigate the relationships between past, present, future, and time. This is because only within STR can definitions of time be formulated which correspond—from the point of view of these relationships—to the appropriate versions of the theory of past, present, and future mentioned above.

Let us finally add that STR is closer to the everyday understanding of time and, as a result, versions of the theory of past, present, and future developed within its framework are closer to commonly accepted notions. The fact that classical mechanics is a limiting case of STR (I do not intend to enter into a discussion of this relationship) serves as proof for the first part of this claim—one that is not without importance. Even today philosophical problems of past, present, and future are often investigated by philosophers without any regard to the work of physicists, relativistic physics included, but simply on the strength of common intuitions. In fact, it is claimed that physics does not provide answers to questions involved here and—what is worse—may even conflict with intuition. In the course of our enquiry into theories of past, present, and future within the STR framework, this belief will be found to be completely erroneous.

The fourth of the initial assumptions adopted in this book bears explicitly upon the manner of defining past, present, and future. These objects are defined by means of the time relations *earlier*, *simultaneously*, and *later*—an assumption that has several important consequences.

To begin with, this assumption implies that such definitions can in fact be constructed and that they are adequate. Many contemporary philosophers of time would, however, disagree with this (see, for instance, Gale, 1968). They would argue that definitions formulated this way are not adequate, and that they allow some important components of meaning which truly characterize the concepts of past, present, and future to be lost 'on the way'. They would insist that it is not possible to define these concepts correctly by means of the relations earlier, simultaneously, and later, while, on the other hand, it is possible, and in fact necessary to describe the latter with the help of the former, i.e., to perform the inverse operation. This view will be discussed in the final chapter.

Another consequence of the adopted assumption is that past, present, and future defined in such a manner have a relational character, i.e., they are always objects related to some particular definite object. What type of object it is depends on the type of set in which the relations earlier, simultaneously, and later are specified. For instance, if the relation later is specified in the set of events, then we speak about the future of a particular event, e.g. x. When defining the future by means of the relation later, therefore we have to frame it in a relational way, i.e., to relate it to a specific object. The same, obviously, concerns past and present defined by means of the relations earlier and simultaneously. All this coincides with current intuition concerning the meaning of the concepts defined.

Nevertheless, here too some objections are raised by the same philosophers who treat past, present, and future in a non-relational absolute way, at any rate in their general statements. This is because they deny that it is possible to arrive at an adequate definitions of these concepts by means of time relations. These differing views are contrasted in the final chapter.

Finally, our assumption has yet another important consequence connected with the adoption of STR as a physical framework of the present discussion. In STR, the relations earlier, simultaneously, and later are defined on the set of events, and also (in a definitionally derivative manner) on the set of moments. In consequence, if we adopt those relations as a basis of the definition of past, present, and future, then at once we obtain two versions of the theory of these objects; the first concerns the past, present, and future of particular events, while the second concerns the past, present, and future of particular moments. Moreover, what is more important, the time relations earlier, simultaneously, and later in STR have a dual character: they are *relative* (dependent on inertial reference systems) and *absolute* (independent of such systems). This is true of time relations specified in the set of events; when specified in the set of moments, they are purely relative. In consequence, we obtain two versions of the theory of past, present, and future, depending on the type of time relations that was used to define those objects: the first tells us about the relative past, present, and future of particular events, while the second tells us about the absolute past, present, and future of particular events.

The fifth and last of the initial assumptions is the hypothesis that the past and future (of events and moments) differ physically from one another. According to our fourth assumption—namely that the past and future are defined by means of the relations earlier and later, respectively—the hypothesis boils down to the postulate that these relations are different from a physical point of view. This latter statement is a formulation of the well-known assumption about the anisotropy of time (cf. Augustynek, 1970). What is meant here, obviously, is the nomological anisotropy, i.e., the claim that at least some laws of physics are not invariant with respect to the transformation of time into itself known as time inversion. What we are talking about, therefore, is the nomological physical difference between the relations earlier and later, and consequently between past and future.

It should be pointed out that, until quite recently, the weaker assumption was accepted in physics, namely the so-called *de facto anisotropy of time* (cf. Augustynek, 1970). At the same time, there was acceptance of a thesis concerning the *nomological isotropy of time*, i.e., the theorem that all laws of physics are invariant with respect to the inversion of time (the principle of *T*-invariance), and that, in consequence, there is no nomological difference between the relations earlier and later, and thus between past and future. After the experiments conducted in Princeton in 1964 (Christenson et al., 1964), concerning neutral kaons, and subsequently repeated many times, physicists came to believe (although with some delay) that some laws of physics regarding weak interactions are asymmetric with respect to time, i.e., that the nomological anisotropy of time is actually a fact (in other words, that the *T*-invariance principle does not hold universally). That is why I consider the adoption of the stronger nomological assumption concerning the physical difference between past and future to be justified.

It is worth noting that a different invariance principle regarding time still holds firmly. What I have in mind is the assumption that all laws of physics are invariant with respect to the tranformation of time into itself, which is known as the *translation of time*. The principle is equivalent to the theorem that all moments of time do not differ nomologically from each other (they are physically equivalent) or, in other words, that time is homogeneous (cf. Augustynek, 1968). This principle, of course, differs from the *T*-invariance principle mentioned above. Thus, time is anisotropic but undoubtedly homogeneous.

To conclude, a few remarks are in place about the two following chapters which present the basic versions of the theory of past, present, and future in the form of axiomatic systems. The first section of each of these chapters consists of axioms concerning the properties of the time relations adopted as primitive. The second section of each chapter consists of the corresponding definitions of past, present, and future based on these time relations. The third sections comprise axioms regarding the concepts defined thus, and the fourth sections present the most important theses implied by both sets of axioms and definitions. In more complex cases, proofs of such theses will be included, otherwise, only the numbers of the necessary axioms, definitions, and theorems are noted in braces on the right-hand side of the thesis.

CHAPTER 2

THE STANDARD THEORY OF PAST, PRESENT, AND FUTURE

It has already been stated that within STR there appear two forms of the time relation *earlier* defined on the set S, and consequently, there are also two forms of the relations *later* and *simultaneously*. The first form is relative—i.e., dependent on the inertial reference system—relation earlier. In other words, not every two events which bear the relation to each other in one reference system will also bear it in another system. The same is true of the relative relations later and simultaneously, defined by means of the relative relation earlier.

Let the symbol **U** denote the set of inertial reference systems, and the symbols $u, u_1, u_2, \ldots$ elements of the set, that is, $u, u_1, u_2, \ldots \in \mathbf{U}$. Then the particular relations earlier assigned respectively to the systems mentioned above will be denoted by W_u, W_{u_1}, W_{u_2}, ... Let us note that the relativity of the relation earlier consists precisely in the fact that for the different reference systems u_1 and u_2 the corresponding relations W_{u_1} and W_{u_2} are different. This does not mean that the relations are mutually exclusive, it appears that they all have a common product, thanks to which the absolute relation earlier exists.

The primitive terms of the axiomatic system of past, present, and future being constructed here (which is called *standard* or *referential*) are: **S**, which denotes the set of events; **T**, which denotes the set of things (they are subsets of the set **S**); **U**, which denotes the set of inertial reference systems (these are some specific things or subsets of the set **S**); and finally W_u, denoting the relation *relatively earlier* defined on the set **S**.

In the set **S**, the relation relatively earlier is asymmetric, transitive, and therefore irreflexive:

(A1) $W_u \in$ **asymm**,

(A2) $W_u \in$ **trans**,

(T1) $W_u \in$ **irrefl.** {A1, A2}

In view of (A1) and (A2), the relation W_u partially orders the set **S**, as we do not postulate its connectivity in **S**.

The further time relations necessary here, i.e., *relatively later* W_u^* and *relatively simultaneously* R_u, are defined by means of the relation relatively earlier W_u:

(D1) $W_u^* \overset{\text{def}}{=} \breve{W}_u,$

(D2) $R_u \overset{\text{def}}{=} \overline{W}_u \cap \overline{\breve{W}}_u.$

Expressed in words: the relation relatively later is a converse of the relation relatively earlier, while relatively simultaneously is a product of the complement of the relation relatively earlier and the complement of the relation relatively later. Note: the relation relatively later will be hereafter denoted by the symbol $\breve{W}_u$ only.

The above relation is also asymmetric, transitive, and therefore irreflexive in the set **S**:

(T2) $\breve{W}_u \in$ **asymm,** {A1, D1}

(T3) $\breve{W}_u \in$ **trans,** {A2, D1}

(T4) $\breve{W}_u \in$ **irrefl.** {T1, D1}

In the set S, the relation of relative simultaneity R_u is symmetric and reflexive, as well as transitive—which has to be additionally postulated:

(T5) $R_u \in$ **symm,** {D2}

(T6) $R_u \in$ **refl,** {D2, T1}

(A3) $R_u \in$ **trans.**

Consequently, the relation R_u is an equivalence relation in the set S or, in other words, is time-equality, which plays an important role in our theory.

It is not out of place to reflect at this point on the matter of operational definitions for the relative time relations W_u, R_u, and $\breve{W}_u$. I have in mind definitions which from the logical point of view have a conditional character, and are physically based on optical signals. Such a definition for the relation of relative simultaneity was, of course, formulated by Einstein and runs as follows (we ignore the general quantifiers of x and y):

$$T(x, y) \rightarrow [R_u(x, y) \equiv Q(x, y)];$$

$T(x, y)$ means here that from points p and q of an inertial reference system at which the events x and y occur light signals are emitted coincidentally with the occurrence of these events; $Q(x, y)$ means that these signals meet at the midpoint of the segment $\overline{pq}$; $R_u(x, y)$ means that the simultaneity relation R_u holds for the events x and y. Thus the relation requires an operational character.

It must be pointed out that—as far as I know—no one has yet succeeded in formulating an analogous (i.e., conditional and signal-based) operational definition for the remaining time relations, namely for the relation relatively earlier W_u, and relatively later $\breve{W}_u$. It seems that there is some inherent fundamental difficulty here.

The essence of the theory of past, present, and future under discussion—as has already been stated—consists in the definitions of these objects by means of the respective time relations W_u, R_u, and $\breve{W}_u$ introduced here. In view of this, these objects have a relational character, i.e., they always refer to definite correlates, in this instance, to events, since the relations W_u, R_u, $\breve{W}_u$ are defined on the set of events S. Accordingly, past, present, and future must bear the indices of event-symbols. Furthermore, because of the relativity of the above time relations, past, present, and future are also relative, and therefore refer to specific inertial systems; they must, therefore, be indexed by symbols of such systems. To resume: in the theory there is a double relativization of past, present, and future with respect both to events, and to systems.

This being so, we introduce the symbols P_x^u, N_x^u, F_x^u which denote, respectively, the past of the event x in the system u, the present of the event x in the system u, the future of the event x in the system u. Consequently, expressions of the form $z \in P_x^u$, $v \in N_x^u$, $t \in F_x^u$ mean respectively: z occurs in the past of x in the system u, v occurs in the present of x in the system u, t occurs in the future of x in the system u. In this connection, the theory being formulated here will be referred to for short as the $P^uN^uF^u$ theory.

The stipulated definitions of the theory are:

(D3) $P_x^u \overset{\text{def}}{=} \{y \in \mathbf{S}\colon W_u(y, x)\}$,

(D4) $N_x^u \overset{\text{def}}{=} \{y \in \mathbf{S}\colon R_u(y, x)\}$,

(D5) $F_x^u \overset{\text{def}}{=} \{y \in \mathbf{S}\colon \breve{W}_u(y, x)\}$.

In words: the past of the event x in the system u (i.e. P_x^u) is the set of all events y which are earlier than the event x in the system u; the present of the event x in the system u (i.e. N_x^u) is the set of all events y which are simultaneous with the event x in the system u; the future of the event x in the system u (i.e. F_x^u) is the set of all events which are later than the event

x in the system u. Thus P_x^u, N_x^u, F_x^u are certain subsets of the set S, or certain sets of events, and as such are contained in the fundamental layer of reality.

Let us now proceed to the first group of theorems about P_x^u, N_x^u, and F_x^u which are based, among other things, on their definitions:

(T7) $\bigwedge_u \bigwedge_x [x \notin P_x^u]$, {D3, T1}

(T8) $\bigwedge_u \bigwedge_x [x \notin F_x^u]$, {D5, T4}

(T9) $\bigwedge_u \bigwedge_x [x \in N_x^u]$. {D4, T6}

That is, events do not occur either in their own past, or in their own future, whereas they do occur in their own present. This indicates a certain additional difference between P_x^u or F_x^u, and N_x^u.

The sets P_x^u, N_x^u, F_x^u are pairwise disjoint:

(T10) $\bigwedge_u \bigwedge_x [P_x^u \cap F_x^u = \emptyset]$ {D3, D5, A1, T2}

(T11) $\bigwedge_u \bigwedge_x [N_x^u \cap P_x^u = \emptyset]$, {D2, D3, D4}

(T12) $\bigwedge_u \bigwedge_x [F_x^u \cap N_x^u = \emptyset]$. {D4, D5, D2}

Furthermore, the sets P_x^u, N_x^u, F_x^u together cover the set S:

(T13) $\bigwedge_u \bigwedge_x [P_x^u \cup N_x^u \cup F_x^u = \mathrm{S}]$. {D3, D4, D5, D2}

Six further axioms will now be introduced covering certain connections between the sets P_x^u, N_x^u, F_x^u, and the sets P_y^u, N_y^u, F_y^u, where x and y are arbitrary events. What we are dealing with are the past, present, and future of different events in the same inertial reference system:

(A4) $\bigwedge_u \bigwedge_x \bigwedge_y \{W_u(x, y) \rightarrow [F_x^u \cap P_y^u \neq \emptyset]\}$,

(A5) $\bigwedge_u \bigwedge_x \bigwedge_y \{W_u(x, y) \rightarrow [F_x^u \cap N_y^u \neq \emptyset]\}$,

(A6) $\bigwedge_u \bigwedge_x \bigwedge_y \{W_u(x, y) \rightarrow [N_x^u \cap P_y^u \neq \emptyset]\}$,

(A7) $\bigwedge_u \bigwedge_x \bigwedge_y \{P_x^u \cap P_y^u \neq \emptyset\}$,

(A8) $\bigwedge_u \bigwedge_x \bigwedge_y \{F_x^u \cap F_y^u \neq \emptyset\}$,

(A9) $\bigwedge_u \bigwedge_x \bigwedge_y \{R_u(x, y) \equiv [N_x^u \cap N_y^u \neq \emptyset]\}$.

Let us note at once that (A4) expresses the fact that the set **S** is dense with respect to the relation W_u (it is also obviously dense with respect to the relation $\breve{W}_u$), since the axiom is equivalent—according to (D3) and (D5)—to the statement:

$$\bigwedge_u \bigwedge_x \bigwedge_y \{W_u(x, y) \rightarrow \bigvee_z [W_u(x, z) \wedge W_u(z, y)]\}.$$

It follows from (A7) and (A8) that the sets P_x^u and F_x^u are non-empty, while the non-emptiness of the set N_x^u follows from (T9):

(T14) $\bigwedge_u \bigwedge_x [P_x^u \neq \emptyset]$, {A7}

(T15) $\bigwedge_u \bigwedge_x [N_x^u \neq \emptyset]$, {T9}

(T16) $\bigwedge_u \bigwedge_x [F_x^u \neq \emptyset]$. {A8}

Let us note that (T14) is equivalent to the statement that in the set **S** there is no first element with respect to the relation W_u, whereas (T16) is equivalent to the statement that in the set **S** there is no last element with respect to this relation. We write these statements as follows:

$$\bigwedge_u \bigwedge_x \bigvee_y [W_u(y, x)],$$

$$\bigwedge_u \bigwedge_x \bigvee_z [W_u(x, z)].$$

In other words: the set **S** has neither a temporal beginning, nor a temporal end. According to what was said in Chapter 1 (part of the second assumption), this is in agreement with the topological assumption of STR that the set **S** does not have boundary points (events) in the time dimension. In fact, STR assumes more—namely that the set **S** is also unbounded in the metric sense in this dimension. Of course, if this were not so, i.e., if the set **S** had either a beginning or an ending event, then the beginning event would not have had a non-empty past, and the ending event would not have had a non-empty future.

We now introduce the further theses of the $P^uN^uF^u$ theory. Firstly, the sets P_x^u, N_x^u, and F_x^u separately do not cover the set **S**, that is, each of them is only a certain proper part of this set:

(T17) $\bigwedge_u \bigwedge_x \bigvee_{y'} [y' \notin P_x^u]$, {T10, T16}

(T18) $\bigwedge_u \bigwedge_x \bigvee_{t'} [t' \notin N_x^u]$, {T11, T14}

(T19) $\bigwedge_u \bigwedge_x \bigvee_{z'} [z' \notin F_x^u]$. {T10, T14}

Secondly, the following self-evident theses are important:

(T20) $\bigwedge_u \bigwedge_x \bigwedge_y [y \in P^u_x \equiv x \in F^u_y]$, {D3, D5, D1}

(T21) $\bigwedge_u \bigwedge_x \bigwedge_t [t \in N^u_x \equiv x \in N^u_t]$, {D4, T5}

(T22) $\bigwedge_u \bigwedge_x \bigwedge_z [z \in F^u_x \equiv x \in P^u_z]$. {D3, D5, D1}

Thirdly, we have

(T23) $\bigwedge_u \bigwedge_x \bigvee_y [x \in P^u_y]$, {T16, T22}

(T24) $\bigwedge_u \bigwedge_x \bigvee_t [x \in N^u_t]$, {T15, T21}

(T25) $\bigwedge_u \bigwedge_x \bigvee_z [x \in F^u_z]$. {T14, T20}

(T23) states that every event occurs in the past of a certain event; (T24) states that every event occurs in the present of a certain event; (T25) states that every event occurs in the future of a certain event. Furthermore, we have

(T26) $\bigwedge_u \bigwedge_x \bigvee_{y'} [x \notin P^u_{y'}]$, {T19, T22}

(T27) $\bigwedge_u \bigwedge_x \bigvee_{t'} [x \notin N^u_{t'}]$, {T18, T21}

(T28) $\bigwedge_u \bigwedge_x \bigvee_{z'} [x \notin F^u_{z'}]$. {T17, T20}

(T26) state that every event occurs not in the past of a certain event; (T27) states that every event occurs not in the present of a certain event; (T28) states that every event occurs not in the future of a certain event.

Let us note that (T23) and (T26) express the relational character of the past, i.e., the fact that the same events occur in the past of a certain event, and they do not occur in the past of some other event (of course, in the same fixed reference system). The theorems (T24), (T27), and (T25), (T28) express the analogous feature of the present and the future.

There is also one additional relativization peculiar to the $P^uN^uF^u$ theory presented here; namely, the dependence of P^u_x, N^u_x, F^u_x on an inertial reference system. We will now focus on this, since we have so far deliberately avoided the subject. Let us assume that u_1 and u_2 are two inertial systems of reference which are in a state of uniform motion with respect to each other. Let us fix on some event, for instance x; then the past of the event x will in general be different in both systems, i.e., $P^{u_1}_x \neq P^{u_2}_x$. Precisely, the same is true of the present and future of the event x in these systems, so one obtains respectively: $N_x^{u_1} \neq N_x^{u_2}$, and $F_x^{u_1} \neq F_x^{u_2}$. This relativity of past, present, and future follows of course from the fact that P_x^u, N_x^u, F_x^u are defined by means of the time relations W_u, R_u, $\breve{W}_u$, which, according to STR, are themselves relative and dependent on an inertial reference system. In the case considered here, we therefore have: $W_{u_1} \neq W_{u_2}$, $R_{u_1} \neq R_{u_2}$, $\breve{W}_{u_1} \neq \breve{W}_{u_2}$.

The striking and paradoxical consequence, from a commonsense point of view, of the above state of affairs is a situation which could be described by the following statement: for an arbitrary event x, there exists an event y and also three different systems u_1, u_2, u_3, so that the event y occurs coincidentally in the past of the event x with respect to u_1, in the present of the event x with respect to u_2, and in the future of the event x with respect to u_3:

(i) $$\bigwedge_x \bigvee_y \bigvee_{u_1, u_2, u_3} [y \in P_x^{u_1} \wedge y \in N_x^{u_2} \wedge y \in F_x^{u_3}].$$

This is not a thesis of our $\mathrm{P^u N^u F^u}$ theory; for that purpose, it would have to be supplemented by new axioms. The thesis is nevertheless almost trivial in the light of STR and the above considerations. Its paradoxical character arises from the surprise caused by the non-exclusiveness of the sets $P_x^{u_1}, N_x^{u_2}, F_x^{u_3}$, which in turn comes from ignoring their relativization to different systems. The sets P_x^u, N_x^u, F_x^u are exclusive for a particular event in one and the same reference system only.

Within the standard theory presented here, one can introduce notions of past, present, and future which—while being system-related—also refer to things (according to eventism, things are treated, of course, as specific sets of events). This, however, involves a definite event occurring in a given thing. The symbols $P_{a|x}^u$, $N_{a|x}^u$, $F_{a|x}^u$ denote: the past of the thing a with respect to the event x in the system u; the present of the thing a with respect to the event x in the system u; the future of the thing a with respect to event x in the system u.

Designating things by a, b, c, etc., we define the sets:

$$P_{a|x}^u \overset{\text{def}}{=} \{y \in a\colon\ W_u(y, x)\},$$

$$N_{a|x}^u \overset{\text{def}}{=} \{y \in a\colon\ R_u(y, x)\},$$

$$F_{a|x}^u \overset{\text{def}}{=} \{y \in a\colon\ \breve{W}_u(y, x)\}.$$

In words: $P_{a|x}^u$ is the set of events which belong to the thing a, and are earlier than the event x; $N^u_{a|x}$ is the set of events which belong to the thing a, and are simultaneous with the event x; $F^u_{a|x}$ is the set of events which belong to the thing a, and are later than the event x. As can be seen, the sets $P^u_{a|x}, N^u_{a|x}, F^u_{a|x}$ consist only of events belonging to a given thing; therefore we are justified in speaking about the past, present, and future of a given thing.

This is the way the past, present, and future of things (e.g. persons) are normally understood in everyday life (if we ignore relativization to a system). In the same way the past, etc. of things is relativized to certain events (states). Consequently, if we have, for instance, a situation where two events x and y, with respect to which the relativization is introduced, bear the relation earlier to one another, then the past of a given thing with respect to x and y differs, that is $P^u_{a|x} \neq P^u_{a|y}$. Let us also add that the following connection takes place: $P^u_{a|x} \subset P^u_{a|y}$. Precisely the same is true of the present and future of things.

Let us note here that $P^u_{a|x}, N^u_{a|x}, F^u_{a|x}$ are—as is implied by their definitions—temporal parts of the thing a, which obviously follow successively in time. This is analogous to P^u_x, N^u_x, F^u_x being temporal parts of the universe of events S, which also succeed in time. Clearly, the sets $P^u_{a|x}$, $N^u_{a|x}, F^u_{a|x}$ are mutually exclusive, and cover the thing a (as the sets P^u_x, N^u_x, F^u_x cover the set S).

It has been quite common for the concepts of past, present, and future to be used not in relation to events or to things but to moments as, for instance, in the work of some philosophers of time. Within the standard outline presented here, one can define these concepts as well, and sketch their theory. Let us now try to do so systematically.

We know (T5, T6, A3) that the relation of simultaneity R_u is an equivalence relation in the set S. This being the case, it divides this set into a family of equivalence classes designated by $S|R_u$. It is these classes, designated by m, n, ..., that are the specific sets of events that actually constitute moments. In the set of moments $S|R_u$, the relation earlier W'_u is defined by means of the relation earlier W_u (cf. Augustynek, 1975); this relation is not only asymmetric and transitive, but also connected in the set $S|R_u$, thus it orders the set linearly. It is this ordered set of moments, that is, the structure $\langle S|R_u, W'_u\rangle$, which will be called time (defined by abstraction). With the help of the relation W'_u, the following relations are defined within the set $S|R_u$: later $\breve{W}'_u$, and simultaneously R'_u; this latter corresponds to the relation of logical identity. Let us add that because R_u is a relative relation of simultaneity, the set $S|R_u$ is also always relativized to a definite inertial reference system. The same, of course, is true of its elements, i.e., moments. The relations W'_u, R'_u, $\breve{W}'_u$ defined on the set $\mathbf{S}|R_u$ are also relative (as follows from their definitions).

We are now in a position to define the concepts mentioned at the beginning. Let the symbols P^u_m, N^u_m, F^u_m denote respectively: the past of the moment m in the system u; the present of the moment m in the system u; the future of the moment m in the system u. Here are the stipulated definitions:

$$P^u_m \overset{\text{def}}{=} \{n \in \mathbf{S}|R_u\colon\ W'_u(n, m)\},$$

$$N^u_m \overset{\text{def}}{=} \{n \in \mathbf{S}|R_u\colon\ R'_u(n, m)\},$$

$$F^u_m \overset{\text{def}}{=} \{n \in \mathbf{S}|R_u\colon\ \breve{W}'_u(n, m)\}.$$

In words: P^u_m is the set of moments earlier than the moment m, N^u_m is the set of moments simultaneous with the moment m, F^u_m is the set of events later than the moment m. Let us note immediately that, in view of the fact that the relation R'_u in the set $\mathbf{S}|R_u$ is the same as the relation of identity, we have the equality $N^u_m = \{m\}$.

From the above definitions one can see that P^u_m, N^u_m, F^u_m are specific sets of moments (N^u_m is a unit set of this kind), and as such also specific sets of sets of events, in contrast to P^u_x, N^u_x, F^u_x—which are specific sets of events. Therefore, these first sets are of logical type 2, while those second sets are of logical type 1. It should be stressed that, as the definitions indicate, the past, present, and future of a moment are relative sets dependent on an inertial reference system, since the relations W'_u, R'_u, $\breve{W}'_u$, by means of which the sets P^u_m, N^u_m, F^u_m are defined, are relative. This relativity, as we have already seen, has a deeper basis in the relativity of moments themselves.

Let us now consider which axioms and theorems characteristic of the sets P^u_x, N^u_x, F^u_x preserve or lose their validity, and which should be replaced or added when we come to the sets P^u_m, N^u_m, F^u_m (i.e., when in these axioms and theorems events are replaced by moments, and the set $\mathbf{S}$ is replaced by the set $\mathbf{S}|R_u$).

The theorems concerning the structural properties of the time relations W'_u, R'_u, $\breve{W}'_u$, which are fundamental to the conception of $\mathrm{P^u_m N^u_m F^u_m}$, are in comparison to the relations W_u, R_u, $\breve{W}_u$ supplemented by two new theses. Firstly, the relations W'_u, and $\breve{W}'_u$ are connected within the set of moments $\mathbf{S}|R_u$, thus they completely (linearly) order this set. Secondly, as a result of this, the relation R'_u—as was pointed out earlier—corresponds to the relation of identity.

Let us now proceed to some theorems about properties of the sets P^u_m, N^u_m, F^u_m. From the axioms, definitions, and theorems accepted so far it follows that moments do not occur in their own past and future, but do occur in their own present (this is analogous to the theses (T7), (T8), (T9)), further, that these sets are mutually exclusive (the analogues of the theses (T10), (T11), (T12)), and that they cover the set $S|R_u$, of which they are parts (the analogue of the thesis (T13)). Where the connections between the sets P^u_m, N^u_m, F^u_m, and sets P^u_n, N^u_n, F^u_n are concerned (where m and n are arbitrary moments), we are confronted with the following situation: there is no reason why we should not accept analogues of the axioms (A4)–(A9), and therefore also their logical consequences in the form of analogues of the theses (T14)–(T16); we assume accordingly that the set of moments $S|R_u$ is dense with respect to the ordering relation W'_u (the analogue of (A4)).

Furthermore, we assume the analogues of the axioms (A5) and (A6). Finally, we also adopt the analogues of the axioms (A7), (A8), and (A9); in the latter, the formula $R_u(x, y)$ is replaced by the formula $Id(m, n)$, where the symbol Id denotes the relation of logical identity. This being the case, we have to acknowledge the analogues of the theorems implied by these and preceding axioms, namely (T14), (T15), and (T16) which state the non-emptiness of the sets P^u_m, N^u_m, and F^u_m. Let us note that the non-emptiness of the set P^u_m is equivalent to the fact that in the set of moments $S|R_u$ there is no first element with respect to the relation W'_u, while the non-emptiness of the set F^u_m is equivalent to the fact that in the set of moments $S|R_u$ there is no last element with respect to the relation $\breve{W}'_u$. In other words, in the set $S|R_u$ there are no boundary moments. If this set, ordered by the relation W'_u, is taken to represent time, then this latter theorem states that time has no boundary moments. This is in agreement with the fourth assumption stipulated in Chapter 1.

Finally, the analogues of the theorems (T17)–(T28) must also be adopted, among them those stating that every moment belongs to the past of some moment while not belonging to the past of some other moment; that every moment belongs to the present of some moment while not belonging to the present of some other moment; and finally that every moment belongs to the future of some moment while not belonging to the future of some other moment—all theorems concerning the so-called relational character of the sets P^u_m, N^u_m, F^u_m.

To recapitulate: if we compare the conception of $\mathrm{P^u_mN^u_mF^u_m}$ with the $\mathrm{P^uN^uF^u}$ theory, then from a formal logical point of view we do not find any vital differences, apart from some addenda regarding the initial relations W'_u, R'_u, $\breve{W}'_u$ (and their consequences). This is evidence of the close affiliation of the past, present, and future related to events and to moments.

*

Before we proceed to a consideration of the consequences arising out of the transition from STR to classical mechanics for the standard theory presented here, let us reflect on a possible system-related theory of the non-present and present. The theory originates either as a result of revoking the postulate concerning the physical difference between past and future, or, quite independently, in a situation when we accept this postulate, but choose to ignore it, since we are solely interested in the concepts of non-present and present. As will become evident later, the $\bar{N}^u N^u$ theory (if we may so call it) could be developed either as a fragment of $P^u N^u F^u$ theory, or as a separate theory. We will proceed with the second possibility because of its greater simplicity.

The primitive terms of this axiomatic system are: firstly, the symbols **S**, **T**, **U** of the sets of events, things, and inertial reference systems, respectively; secondly, the symbol $\bar{R}_u$ denoting the time relation of non-simultaneity or time-separation. Clearly, the last relation is relative (as indicated by the letter u), and determined within the set **S**. Let us immediately note that if our theory were treated as a fragment of $P^u N^u F^u$ theory, then the relation $\bar{R}_u$ would be defined as the sum $W_u \cup \breve{W}_u$, that is, as the complement of the relation R_u.

We assume that the relation $\bar{R}_u$ is irreflexive, symmetric, and neither transitive nor intransitive in the set **S**:

(A1′) $\bar{R}_u \in \mathbf{irrefl}$,

(A2′) $\bar{R}_u \in \mathbf{symm}$,

(A3′) $\bar{R}_u \notin \mathbf{trans}$,

(A4′) $\bar{R}_u \notin \mathbf{intrans}$.

The second time relation necessary here, also specified in the set **S**, is the relation of relative simultaneity R^*_u. This is defined by means of the relation $\bar{R}_u$ as its complement (the defined relation will still be symbolized by R_u):

(D1′) $R^*_u \overset{\text{def}}{=} R_u$.

In the set S, the relation R_u is reflexive, symmetric, and—it must be additionally stipulated—also transitive:

(T1') $R_u \in \mathbf{refl}$, {D1', A1'}

(T2') $R_u \in \mathbf{symm}$, {D1', A2'}

(A5') $R_u \in \mathbf{trans}$.

Accordingly, as in $P^u N^u F^u$ theory, the relation of the relative simultaneity R_u is an equivalence relation, and therefore time-equality; whereas the relation of relative time separation $\overline{R}_u$, being the complement of the former relation, is a time-difference. It is easy to establish that $Id \subset R_u$, where Id is the relation of identity. This being the case we get $\overline{R}_u \subset \overline{Id}$, that is, the time-difference is included in the logical non-identity.

Let us note that for the relation of relative simultaneity R_u we adopt Einstein's conditional and signal-based operational definition, as we did in the $P^u N^u F^u$ theory. For the relation of relative time-separation $\overline{R}_u$, we assume an operational definition which arises from the definition of R_u by the following substitutions: the formula $R_u(x, y)$ is replaced by $\overline{R}_u(x, y)$, and $Q(x, y)$, i.e., the statement about the meeting of signals in the middle point is replaced by $\sim Q(x, y)$, i.e., the statement about the absence of such a meeting. By this means one gives an operational meaning to both time relations R_u and $\overline{R}_u$.

We define non-present (for lack of a better term) and present by analogy as before by means of the relations $\overline{R}_u$ and R_u, respectively. In this connection, those objects have a relational character, i.e., they are always related to some particular events (the relations $\overline{R}_u$ and R_u are defined in S). In this connection too, those objects are relative, i.e., dependent on an inertial reference system (the relations $\overline{R}_u$ and R_u are explicitly relative). This being the case, we introduce the symbols $\overline{N}_x^u$ and N_x^u which denote, respectively, the non-present of the event x in the system u, and the present of the event x in the system u. It will be readily noted that if the theory being described is treated as a fragment of $P^u N^u F^u$ theory, then the non-present $\overline{N}_x^u$ is simply the sum of the past P_x^u and future F_x^u.

The stipulated definitions are as follows:

(D2') $\overline{N}_x^u \overset{\text{def}}{=} \{y \in S: \overline{R}_u(y, x)\}$,

(D3') $N_x^u \overset{\text{def}}{=} \{y \in S: R_u(y, x)\}$.

In words: the non-present of the event x in the system u, i.e., $\overline{N}_x^u$, is the set of all events y which are separated in time from the event x in the system

u; the present of the event x in the system u, i.e., N^u_x, is the set of all events y which are simultaneous with the event x in the system u. Therefore, the sets $\bar{N}^u_x$ and N^u_x are subsets of the set S, and, consequently, specific sets of events.

Now we will outline groups of axioms and theorems corresponding to the appropriate groups of theses from P^{u}N^{u}F^u theory. Firstly, we have:

(T3′) $\bigwedge_u \bigwedge_x [x \notin \bar{N}^u_x]$, {D2′, A1′}

(T4′) $\bigwedge_u \bigwedge_x [x \in N^u_x]$. {D3′, T1′}

This means that an event does not occur in its own non-present, while it occurs in its own present; this expresses an important difference between the present N^u_x and the non-present $\bar{N}^u_x$. Secondly, the sets $\bar{N}^u_x$ and N^u_x are disjoint and, obviously, cover the set S:

(T5′) $\bigwedge_u \bigwedge_x [\bar{N}^u_x \cap N^u_x = \emptyset]$, {D2′, D3′, D1′}

(T6′) $\bigwedge_u \bigwedge_x [\bar{N}^u_x \cup N^u_x = \mathrm{S}]$. {D2′, D3′, D1′}

Let us consider possible connections between the sets $\bar{N}^u_x$, N^u_x, and $\bar{N}^u_y$, N^u_y, where x and y are some arbitrary events (obviously, assuming the same reference system). Here are the connections:

(A6′) $\bigwedge_u \bigwedge_x \bigwedge_y [\bar{N}^u_x \cap \bar{N}^u_y \neq \emptyset]$,

(A7′) $\bigwedge_u \bigwedge_x \bigwedge_y [\bar{R}_u(x, y) \equiv \bar{N}^u_x \cap N^u_y \neq \emptyset]$,

(A8′) $\bigwedge_u \bigwedge_x \bigwedge_y [R_u(x, y) \equiv N^u_x \cap N^u_y \neq \emptyset]$.

The non-presents of two arbitrary events x and y therefore always have a non-empty product (A6′). However, their presents have such a non-empty product only when the events x and y are simultaneous (A8′). The product of the non-present of x and present of y is non-empty only when the events are separated in time (non-simultaneous) (A7′).

From (A6′) it follows that the non-present of the event x, i.e., $\bar{N}^u_x$, is non-empty, while from (T4′) it follows that the present of the event x, i.e., N^u_x, is non-empty:

(T7′) $\bigwedge_u \bigwedge_x [\bar{N}^u_x \neq \emptyset]$, {A6′}

(T8′) $\bigwedge_u \bigwedge_x [N^u_x \neq \emptyset]$. {T4′}

Now we may formulate further theorems of the $\bar{N}^u N^u$ theory. Firstly, the sets $\bar{N}_x^u$ and N_x^u do not cover separately the set S, which means that they are certain proper parts of this set:

$$(T9') \quad \bigwedge_u \bigwedge_x \bigvee_{y'} [y' \notin \bar{N}_x^u], \qquad \{T5', T8'\}$$

$$(T10') \quad \bigwedge_u \bigwedge_x \bigvee_{z'} [z' \notin N_x^u]. \qquad \{T5', T7'\}$$

Secondly, the following theses are important:

$$(T11') \quad \bigwedge_u \bigwedge_x \bigwedge_y [y \in \bar{N}_x^u \equiv x \in \bar{N}_y^u], \qquad \{D2', A2'\}$$

$$(T12') \quad \bigwedge_u \bigwedge_x \bigwedge_y [y \in N_x^u \equiv x \in N_y^u]. \qquad \{D3', T2'\}$$

Thirdly, we have:

$$(T13') \quad \bigwedge_u \bigwedge_x \bigvee_y [x \in \bar{N}_y^u], \qquad \{T7', T11'\}$$

$$(T14') \quad \bigwedge_u \bigwedge_x \bigvee_z [x \in N_z^u]. \qquad \{T8', T12'\}$$

(T13′) states that every event occurs in the non-present of a certain event, and (T14′) states that every event occurs in the present of a certain event. Finally, we obtain

$$(T15') \quad \bigwedge_u \bigwedge_x \bigvee_{y'} [x \notin \bar{N}_{y'}^u], \qquad \{T9', T11'\}$$

$$(T16') \quad \bigwedge_u \bigwedge_x \bigvee_{z'} [x \notin N_{z'}^u]. \qquad \{T10', T12'\}$$

(T15′) states that every event does not occur in the non-present of a certain event; (T16′) states that every event does not occur in the present of a certain event.

The theorems (T13′) and (T15′) together express the relativity of non-present, that is, the fact that the same events occur in the non-present of some event, and do not occur in the non-present of some other event (in the same reference system). The relativity of present is expressed by the theorems (T14′) and (T16′) together.

In the $\bar{N}^u N^u$ theory being developed there is, of course, also the dependence of $\bar{N}_x^u$ and N_x^u on an inertial reference system which so far has been deliberately left out. Let u_1 and u_2 be inertial reference systems moving with respect to each other. Then the non-present and present of some specific event, e.g. x, with respect to u_1 and u_2 will generally be different, i.e., $\bar{N}_x^{u_1} \neq \bar{N}_x^{u_2}$, and $N_x^{u_1} \neq N_x^{u_2}$. This relativity of $\bar{N}_x^u$ and N_x^u follows, as is well known, from the fact that they are defined by means of the time relations $\overline{R}_u$ and R_u. According to STR, these last relations are relative and dependent on an inertial reference system; thus we have in this case $\overline{R}_{u_1} \neq \overline{R}_{u_2}$, and $R_{u_1} \neq R_{u_2}$.

Within the framework of $\overline{N}^u N^u$ theory, we may introduce system-related concepts of non-present and present referring to things or, more strictly, to events occurring in things. The non-present of the thing *a* with respect to the event *x* in the system *u* can be defined by means of the relation $\overline{R}_u$; the present of the thing *a* with respect to the event *x* in the system *u* can be defined by means of the relation R_u.

Finally, within $\overline{N}^u N^u$ theory, we may introduce system-related concepts of non-present and present referring to moments. The relations $\overline{R}'_u$ and R'_u may be defined in the set of moments $S|R_u$ by means of the relations $\overline{R}_u$ and R_u. The former ($\overline{R}'_u$) is the relative time-separation for moments, the latter (R'_u) we know to be the relation of identity for moments. By means of these relations, we can define the non-present of the moment *m* in the system *u*, and the present of the moment *m* in the system *u*. These will be specific sets of moments—non-empty, exclusive, and covering the set $S|R_u$. Let us note that there is only one possible definition of time in this conceptual framework, namely the structure $\langle S|R_u, \overline{R}'_u\rangle$, where $\overline{R}'_u$ does not order the set $S|R_u$.

In conclusion, it is worth having a brief look at the relative space relations analogous to the time relations $\overline{R}_u$ and R_u, since they will be of use later. We refer here to the relative space-separation relation $\overline{L}_u$, and to the relation of relative co-location L_u which is defined as the complement of $\overline{L}_u$. Those relations are obviously defined on the set of events S. It should be pointed out that, as far as we know, no one has yet succeeded in constructing operational definitions of those relations, analogous to the operational definitions of the relations $\overline{R}_u$ and R_u. Within STR it is assumed that the relation of relative space-separation $\overline{L}_u$ is irreflexive and symmetric, and that it is neither non-transitive, nor transitive in the set S. Therefore, it is a relation of spatial difference. From this it follows that the relation of relative co-location L_u is reflexive and symmetric in the set S; it is also assumed that it is transitive. Thus it is a relation of spatial equality.

*

Let us now consider the implications of the limit-transition from STR to classical mechanics (CM) for the standard $P^u N^u F^u$ theory. We will also tackle the connection between the concepts of past, present, and future within CM on one hand, and within the framework of the commonsense view on the other.

In the standard theory, the past, present, and future of an event are relative, i.e., dependent on some inertial reference system. In other words, in different inertial systems we have, in principle, the following differences: $P_x^{u_1} \neq P_x^{u_2}, N_x^{u_1} \neq N_x^{u_2}, F_x^{u_1} \neq F_x^{u_2}$. This relativity, as is known, follows from the fact that the sets P_x^u, N_x^u, F_x^u are defined by means of the time relations $W_u, R_u, \breve{W}_u$; and also from the fact that these relations are relative (i.e., dependent on an inertial reference system), that is, we have, in principle, the following differences: $W_{u_1} \neq W_{u_2}, R_{u_1} \neq R_{u_2}, \breve{W}_{u_1} \neq \breve{W}_{u_2}$. Let us add, parenthetically, that the relativity of these relations implies the relativity of time defined by abstraction (by means of the relation R_u), or defined relationally (by means of the relation W_u only)—different times are related to different systems.

The contemporary version of CM is the limiting case of STR. This means the following: STR is characterized by the spatio-temporal Lorentz transformations, while CM is characterized by the temporal and spatial transformations of Galileo. In certain physical conditions, the Galilean transformations are logically implied by Lorentz transformations, which means that CM follows from STR. It is precisely in this sense that one speaks of CM being the limiting case of STR. The conditions are expressed by the formula $v \ll c$, where v designates the relative velocity of different inertial systems, while c designates the velocity of light in a vacuum, which represents the maximum velocity of physical interactions in nature. The formula states that v is very small in comparison to c. Consequently, STR implies CM in such a subset of inertial systems in which every two systems belonging to it move with respect to one another with a velocity v fulfilling the condition $v \ll c$. Let us note that such velocities are generally characteristic of macro-bodies.

According to Galilean transformations (CM), the time relations earlier, simultaneously, and later are not relative but absolute, i.e., independent from an inertial reference system. From the point of view of the limit-transition from STR to CM described above, this takes place only if the condition $v \ll c$ is realized; in other words, in the subset of inertial systems specified above where the condition is satisfied. One should nevertheless note that time relations in such conditions are absolute only in an approximate manner, with a high degree of approximation. This arises from the fuzzy character of the transition condition $v \ll c$. CM applies quite accurately only if $v = 0$.

Under the circumstances described, the relations W_u, R_u, $\breve{W}_u$ are absolute; thus the sets P_x^u, N_x^u, F_x^u defined by means of them are also (approximately) absolute. Consequently, for different elements u_1 and u_2 of the subset U' of inertial systems, satisfying the transition condition, we will have the following connections: $P_x^{u_1} = P_x^{u_2}$, $N_x^{u_1} = N_x^{u_2}$, $F_x^{u_1} = F_x^{u_2}$. This means that under the specified circumstances the past, present, and future of a given event will not be different in various inertial systems. We therefore have one meta-systemic past of a given event, one meta-systemic present of the event, and, finally, one meta-systemic future of the event. As a result of this, the paradoxical consequence of the relativity of P_x^u, N_x^u, F_x^u, according to which a given event can occur simultaneously in the past, present, and future of some other event (i.e., in different reference systems), described by the theorem (i), is eliminated. It is worth mentioning that under these circumstances there is only one (meta-systemic) time, defined by abstraction or relationally; therefore, time is (approximately) absolute.

Everything that has been said here about the sets P_x^u, N_x^u, F_x^u obviously also applies to the non-present and present of events, i.e., to the sets $\bar{N}_x^u$ and N_x^u. This means that in the set of systems in which relative velocities satisfy the condition $v \ll c$, the sets $\bar{N}_x^u$ and N_x^u are absolute, i.e., independent from those inertial systems, because under these circumstances the relations of time-separation and of simultaneity, by means of which the above sets are defined, are absolute. Clearly, in these conditions, the paradoxical consequence following from the relativity of $\bar{N}_x^u$ and N_x^u, according to which a given event can occur at the same time in the non-present and present of some other event (i.e., in different reference systems), disappears at once.

In CM (if one ignores the fact that it is a limiting case of STR), the time relations W_u, R_u, $\breve{W}_u$, as well as $\bar{R}_u$, are universally absolute, i.e., they are absolute in the whole set of inertial reference systems. This was the general view before the emergence of STR. Consequently, the sets P_x^u, N_x^u, F_x^u were once considered as totally absolute and identical, independent of any conditions whatsoever. The commonsense point of view was precisely the same, i.e., it corresponded completely to CM. It can hardly be wrong to suggest that this correspondence was in fact one of the sources of CM. In turn, CM supported this view and helped to raise it to the ranks of scientifically justified views.

Today, when looked at from the perspective of STR, particularly knowing that CM is its limiting case, this coincidence of views is immediately explicable. Both everyday and scientific experience—the latter until almost the end of the nineteenth century, the former until now—dealt with the motion of macro-bodies (and so with the motion of inertial reference systems) which completely satisfied the condition $v \ll c$. This is the source of the agreement noted above and of the elevation of these views to the status of universal truths, as expressed by Galilean transformations characteristic for CM. On the other hand, STR cancels out this second component of the analysed position, and also of CM, namely the claim to universality. It points out clearly that a transition beyond the condition $v \ll c$, i.e., when the velocities of the reference systems concerned are not too far from c, implies the essential relativity of the time relations W_u, R_u, $\breve{W}_u$, and the sets P^u_x, N^u_x, and F^u_x. Everyday experience, however, still remains within the framework of the condition $v \ll c$. That is why even today the relativity of these relations smacks of a paradox, or the flouting of commonsense. Will this ultimately change? I think it will.

To avoid possible misunderstandings, it seems appropriate here to add two comments to what has been said above. Firstly, one should not identify the absolutization of the time relations W_u, R_u, $\breve{W}_u$, and also the sets P^u_x, N^u_x, F^u_x (which takes place in the course of the transition from STR to CM, when $v \ll c$) with the fact that within STR itself there are absolute time relations earlier, simultaneously, and later. We have already mentioned these relations which, as we shall see later, will constitute the basis of the non-standard theory of past, present, and future. These relations, as will become evident, are genuinely and universally absolute, i.e., independent of inertial reference systems, not only in some specific circumstances like relative time relations, but also in all possible physical conditions.

Secondly, one should not suppose that the conditional absolutization of the time relations W_u, R_u, $\breve{W}_u$, and of the sets P^u_x, N^u_x, F^u_x, described here, implies the conditional elimination of reference, i.e., of the relativization of those relations and sets to specified events. There is no such implication; the last relativization still holds, even in the conditions described above, and has nothing to do with the relativization of these sets to definite systems. Let us add that referring the past, present, and future to events or things is in complete agreement with commonsense.

CHAPTER 3

THE NON-STANDARD THEORY OF PAST, PRESENT, AND FUTURE

As has already been mentioned, in addition to the system-dependent time relations earlier, later and simultaneously, the Special Theory of Relativity (STR) also contains the absolute time relations earlier, later and *quasi-simultaneously*. Analogously to the system-dependent relations, the absolute relations will be defined on the set S. The fact that the relation earlier is an absolute one means that it is independent of an inertial reference system. In other words, every two events which stand in this relation in a certain system stand in it in any other system. The same applies, obviously, to the absolute relations later and quasi-simultaneously. The relations will be denoted by W, R, $\breve{W}$, respectively; the third relation, i.e., absolutely later, is denoted by $\breve{W}$ since in the constructed theory it is defined as the converse of the relation W.

The precise meaning of these relations can be seen through the connections they have with the respective system-dependent time relations. These connections can be easily established within the framework of STR. The following should be noted before discussing these relations; the system-dependent relations denoted by W_u, R_u, $\breve{W}_u$ are actually schemata of different relations dependent on specific reference systems. At the same time, the absolute relations—by virtue of being absolute—are not schemata; lack of the index u indicates lack of relativization.

Here, then, is the relationship between the relation W and the relation W_u and, as a result, between the relations $\breve{W}$ and $\breve{W}_u$, as well as between R and R_u. The following two theorems can be proved according to STR. First, that the relation W is a joint product of all the relations W_u, W_{u_1}, W_{u_2}, ..., characteristic of the respective inertial reference systems. This and the definitions of $\breve{W}$ and $\breve{W}_u$ (as converses of the corresponding relations W and W_u) entails that the relation $\breve{W}$ is a product of all the relations $\breve{W}_u, \breve{W}_{u_1}, \breve{W}_{u_2}, \ldots$ Second, this theorem and the definitions of R and R_u (the relation R is defined as a product $\overline{W} \cap \overline{\breve{W}}$) entails that the relation R is a sum of all the relations $R_u, R_{u_1}, R_{u_2}, \ldots$, characteristic of the respective inertial reference systems.

The first theorem directly implies that the relation *absolutely earlier*, W, is a proper subrelation of the relation relatively earlier, W_u; more precisely, the subrelation of each of the relations $W_u, W_{u_1}, W_{u_2}, \ldots$, i.e., we have $W \subset W_u$, $W \subset W_{u_1}$, etc. Next, this theorem implies that the relation *absolutely later*, $\breve{W}$, is also the proper subrelation of the relation relatively later, $\breve{W}_u$. The second theorem implies that the relation of simultaneity R_u is the proper subrelation of the quasi-simultaneity relation R, i.e., $R_u \subset R$, $R_{u_1} \subset R$, etc. The above situation illustrates, among other things, the logical connections between the absolute and system-dependent relations. This is in compliance with the assumptions concerning the relationship between absolute time relations and the causal relation (or rather, the possibility of its occurrence). These assumptions, also based on STR, will subsequently be formulated and accepted as axioms.

The axiomatic system of past, present, and future, which is built here on the basis of the relations W, R and $\breve{W}$, will be called *non-standard*, or *system independent*, or, for short, *PNF theory*. Absolute time relations are, as a matter of fact, rarely mentioned within STR. The primitives of this system are: **S**, the set of all events; **T**, the set of things (the set of sets of events), W, the relation absolutely earlier defined on **S**, and H, the causal relation also defined on the set **S**.

Two comments should be made on the list of primitive terms. First, unlike the case of $P^uN^uF^u$ theory, the term **U** denoting the set of reference systems does not appear on this list. This is obvious in light of the fact that only absolute time relations are used, and only absolute objects are defined by means of them. Second, given the connections between the absolute and system-dependent time relations presented here, the question arises whether there is any point in constructing an autonomous PNF theory. These connections may be treated as definitions of absolute time relations by means of the corresponding system-dependent relations. In this way, PNF theory can be constructed as a part of $P^uN^uF^u$ theory, where the system-dependent time relations are primitive. However, this will not be done here because we want to demonstrate that PNF theory can be constructed as an independent system, not explicitly connected with the notion of a reference system.

The first group of axioms and theorems of PNF theory expresses the structural properties of the absolute relations W, R, $\breve{W}$. The relation W is asymmetric and transitive in the set S, and is therefore irreflexive:

(A1) $W \in$ **asymm**,

(A2) $W \in$ **trans**,

(T1) $W \in$ **irrefl.** {A1, A2}

Hence, according to (A1) and (A2), the relation W partially orders the set S. It is a partial ordering because we do not assume that the relation W is connected in S.

As has already been mentioned, the relation absolutely earlier W is used to define the relation absolutely later W^* (which is denoted by this symbol only for the sake of formulating its definition) as well as by the absolute relation of quasi-simultaneity R:

(D1) $W^* \overset{\text{def}}{=} \breve{W}$,

(D2) $R \overset{\text{def}}{=} \overline{W} \cap \overline{\breve{W}}$.

The relation $\breve{W}$ is also asymmetric and transitive in the set S, and is therefore irreflexive:

(T2) $\breve{W} \in$ **asymm**, {A1, D1}

(T3) $\breve{W} \in$ **trans**, {A2, D1}

(T4) $\breve{W} \in$ **irrefl.** {T1, D1}

The quasi-simultaneity relation R is symmetric and reflexive in the set S. It is, however, neither transitive nor intransitive there:

(T5) $R \in$ **symm**, {D2}

(T6) $R \in$ **refl**, {D2, T1}

(A3) $R \notin$ **trans**,

(A4) $R \notin$ **intrans.**

Therefore, the relation R is a relation of time-similarity in the set S. That is why it is called quasi-simultaneity in opposition to the relation of simultaneity which is an equivalence and the reforea time-equality. The latter is obviously 'stronger' than time-similarity. Consequently, the relation of simultaneity R_u is contained in the quasi-simultaneity relation R, which is in complete agreement with the connection between these relation as established above.

The problem of the operational definitions of the time relations W, R, $\breve{W}$ will now be discussed. What is meant here are the conditional and signal definitions (see Chapter 2), as in the case of the system-dependent relations W_u, R_u, $\breve{W}_u$. A definition of this kind can be easily formulated for the relation of quasi-simultaneity R. It is as follows: $T(x, y) \rightarrow [R(x, y) \equiv \equiv Q'(x, y)]$. $T(x, y)$ means here that from points p and q (of a definite inertial reference system), at which the events x and y occur, light signals are emitted simultaneously with the occurrence of these events; $Q'(x, y)$ means that these signals meet at any point lying between the points p and q and different from each of them; $R(x, y)$ means that the events x and y are quasi-simultaneous. When this operational definition of the relation R is compared with the operational definition of the relation R_u, the fact that R_u is contained in R is immediately apparent. It should also be noted that this definition of R assumes a certain inertial reference system, since spatial points and distances between them are mentioned. This obviously cannot be otherwise; in order to establish the occurrence of any fact, including the occurrence of the relation R, a definite reference system is required as a practical basis. It does not follow from this, though, that because of this assumption these facts are dependent on inertial reference systems.

Unfortunately, we are not able to formulate the analogous, i.e., conditional and signal, operational definition of the relation absolutely earlier W, and, by the same token, of the relation absolutely later $\breve{W}$. As far as I know, nobody else has succeeded in doing this either. An essential difficulty arises here as in the case of the system-dependent relations W_u and $\breve{W}_u$ (see Chapter 2). Because of the close connections between the system-dependent relations W_u and $\breve{W}_u$, and the absolute relations W and $\breve{W}$, this difficulty seems to be identical. This problem deserves serious investigation since, although the operational meaning of the relations R_u and R is known, that of W_u and $\breve{W}_u$ (as well as that of W and $\breve{W}$) is not, and these are the relations by means of which the relations R_u and R are defined (in a theoretical, i.e., non-operational way, and by equivalence, i.e., non-conditional definition).

As in the case of $P^uN^uF^u$ theory, the essence of the non-standard system discussed here, i.e., PNF theory, consists of the definitions of past, present, and future by means of the corresponding absolute time relations W, R, $\breve{W}$. This is undoubtedly a new idea, at least in the form developed here. The adoption of these definitions implies two consequences. First, past, present, and future have a relational character, i.e., they refer to certain events (or things). Second, since these time relations are absolute, then past, present, and future are absolute as well, i.e., they are independent of any inertial reference system. That is why the symbols of these objects bear indices of events only, and not those of systems.

It is therefore necessary to introduce the symbols P_x, N_x, F_x which denote respectively: the absolute past of the event x, the absolute present of the event x (more precisely, its quasi-present), and the absolute future of the event x. The term 'absolute' will not be used further below (except in definitions) unless the context calls for it. These symbols are entirely sufficient to distinguish these absolute objects from the system-dependent past, present, and future, i.e., from P_x^u, N_x^u, F_x^u. The term 'quasi-present' will not be used either, even though the specific character of the theory under discussion hinges on this point; in this case, the symbol N_x also suffices to distinguish it from N_x^u. According to this notation, the formulae $z \in P_x, v \in N_x, t \in F_x$ mean, respectively, that z occurs in the past of x, v occurs in the present of x and t occurs in the future of x.

The stipulated definitions take the following form:

(D3) $P_x \overset{\text{def}}{=} \{y \in \mathbf{S}\colon\ W(y, x)\}$,

(D4) $N_x \overset{\text{def}}{=} \{y \in \mathbf{S}\colon\ R(y, x)\}$,

(D5) $F_x \overset{\text{def}}{=} \{y \in \mathbf{S}\colon\ \breve{W}(y, x)\}$.

In words: the absolute past of the event x, i.e., P_x, is the set of all events y absolutely earlier than x; the absolute present of the event x, i.e., N_x, is the set of all events y quasi-simultaneous with x; the absolute future of the event x, i.e., F_x, is the set of all events y absolutely later than x. As one can see, P_x, N_x, F_x are proper subsets of the set S, i.e., certain sets of events. They are therefore contained in a fundamental layer of reality.

The defined sets differ from sets P_x^u, N_x^u, F_x^u which are relative and dependent on an inertial reference system. Nevertheless, there are definite connections with the latter, which are a consequence of the connections between the absolute relations $W, R, \breve{W}$ and system-dependent relations $W_u, R_u, \breve{W}_u$, and the definitions (introduced above) of the sets P_x, N_x, F_x. These connections are the following: $P_x \subset P_x^u, N_x^u \subset N_x$ and $F_x \subset F_x^u$. As one can see, absolute past and absolute future are subsets of system-related past and future. Absolute present, i.e., quasi-present, however, is a covering set of the system-related present, which corresponds to the fact that the relation R is a covering relation of the relation R_u.

According to STR, for every event there exists in the set S a light cone which is characteristic for this event. Such a cone is generated by all possible world-lines of photons which pass through this event. Let us consider some event x and its light cone $C(x)$. Let α_1 be the lower internal part of the cone, α_2 the upper internal part of the cone, β the region external to the cone, γ_1 the lower surface of the cone, γ_2 the upper surface of the cone, and $\{x\}$ the set containing the event x only; as can be seen, the sets α_1, $\alpha_2, \beta, \gamma_1, \gamma_2$ do not contain the set $\{x\}$. The cone $C(x)$ is the sum of the sets $\alpha_1, \alpha_2, \gamma_1, \gamma_2$, and $\{x\}$, i.e., $C(x) = \alpha_1 \cup \alpha_2 \cup \gamma_1 \cup \gamma_2 \cup \{x\}$, whereas the set S is the sum of $C(x)$ and the set β external to the cone, i.e., $\mathrm{S} = C(x) \cup \beta$. So $P_x = \alpha_1 \cup \gamma_1$, i.e., it is the lower internal region of the cone $C(x)$ together with its surface (without the event x). $F_x = \alpha_2 \cup \gamma_2$, i.e., it is the upper internal region of the cone together with its surface (without the event x). Finally, $N_x = \beta \cup \{x\}$, i.e., it is the region external to the cone together with the event x. This is a physico-geometrical interpretation of the sets P_x, N_x, F_x.

Here is the first group of theses of our system which apply to the sets P_x, N_x, F_x:

(T7) $\bigwedge_x [x \notin P_x]$, {D3, T1}

(T8) $\bigwedge_x [x \notin F_x]$, {D5, T4}

(T9) $\bigwedge_x [x \in N_x]$. {D4, T6}

Thus an event occurs neither in its past nor in its future, whereas it occurs in its present; this reflects, obviously, a definite difference between the sets P_x and F_x on the one hand, and N_x on another. The sets P_x, N_x, F_x are pairwise disjoint:

(T10) $\bigwedge_x [P_x \cap F_x = \emptyset]$, {D3, D5, A1, T2}

(T11) $\bigwedge_x [N_x \cap P_x = \emptyset]$, {D4, D3, D2}

(T12) $\bigwedge_x [F_x \cap N_x = \emptyset]$. {D4, D5, D2}

Finally, the sets P_x, N_x, F_x cover the set S:

(T13) $\bigwedge_x [P_x \cup N_x \cup F_x = \mathrm{S}]$. {D3, D4, D5, D2}

Let us note that theorems (T7)–(T13) correspond exactly to the interpretations concerning the relationships between the sets P_x, N_x, F_x and the respective regions of the light cone of the event x.

Some interesting connections hold between the past, present, and future of a certain event x, i.e., P_x, N_x, F_x, and the past, present, and future of the event y, i.e., P_y, N_y, F_y, where x and y are arbitrary events. Six further axioms refer to these connections. These axioms can be easily verified by considering the corresponding light cones characteristic of the events x and y. The following condition, provided by STR, should thus be obeyed: the indicated light cones should have the same time orientation. This condition is a consequence of one of the principles of STR which claims that the velocity of light is independent of an inertial reference system, i.e., it is absolute. Here are the proposed axioms:

(A5) $\bigwedge_x \bigwedge_y \{W(x, y) \rightarrow [F_x \cap P_y \neq \emptyset]\}$,

(A6) $\bigwedge_x \bigwedge_y \{[R(x, y) \wedge x \neq y \vee W(x, y)] \rightarrow [F_x \cap N_y \neq \emptyset]\}$,

(A7) $\bigwedge_x \bigwedge_y [R(x, y) \wedge x \neq y \vee W(x, y)] \rightarrow [N_x \cap P_y \neq \emptyset]\}$,

(A8) $\bigwedge_x \bigwedge_y [P_x \cap P_y \neq \emptyset]$,

(A9) $\bigwedge_x \bigwedge_y [F_x \cap F_y \neq \emptyset]$,

(A10) $\bigwedge_x \bigwedge_y [N_x \cap N_y \neq \emptyset]$.

Let us first note that (A5) expresses the fact that the set of events S is dense with respect to the relation absolutely earlier W. This stems from the fact that, according to (D3) and (D5), the axiom (A5) is equivalent to the theorem:

$$\bigwedge_x \bigwedge_y \{W(x, y) \rightarrow \bigvee_z [W(x, z) \wedge W(z, y)]\}.$$

The last three axioms immediately imply that the sets P_x, N_x, F_x are non-empty:

(T14) $\bigwedge_x [P_x \neq \emptyset]$, {A8}

(T15) $\bigwedge_x [F_x \neq \emptyset]$, {A9}

(T16) $\bigwedge_x [N_x \neq \emptyset]$. {A10}

Let us note that (T14) is equivalent to the theorem which states that in the set of events S there is no first element with respect to the relation W, while (T15) is equivalent to the thesis which states that in the set S there is no last element with respect to the relation W. These theorems are as follows:

$$\bigwedge_x \bigvee_y [W(y, x)],$$

$$\bigwedge_x \bigvee_z [W(x, z)].$$

In other words: the universe of events S has neither a temporal begining nor a temporal end, which corresponds to the assumption of STR that the set S does not have boundary points (events) in the time dimension. Theorems analogous to (T14) and (T15) appear, as we have already seen in $P^uN^uF^u$ theory.

The axiom (A10) implies the paradox which is called the *paradox of present*. This paradox is unique for PNF theory (it does not appear in $P^uN^uF^u$ theory) and is expressed by the following thesis:

$$\text{(T17)} \quad \bigwedge_x \bigwedge_y \bigvee_z \{W(x, y) \rightarrow [x \in N_z \wedge y \in N_z]\}. \qquad \{\text{A10, D4}\}$$

In words: for any two events x and y, there exists such an event z that, if x is absolutely earlier than y, then both these events occur in the present of the event z; in other words, they are quasi-simultaneous with z. (A10) together with (D4) also imply the thesis which differs from (T17) only by the fact that in its antecedent $\breve{W}(x, y)$ is substituted for $W(y, x)$.

Two comments may be made here. First, this paradox—as well as the axiom (A10)—is comprehensible, and even obvious, when (A3), according to which the relation of quasi-simultaneity R is not transitive, is taken into consideration. In this case, it is possible that $R(x, z) \wedge R(z, y) \wedge \sim R(x, y)$, and that is what happens here. Second, it is not difficult to note that the larger the temporal distance between the events x and y the large should be the spatial distances between them and the event z. Hence, (T17) and obviously also (A10) assume that the set of events S is unbounded in a metric sense in its spatial dimensions. Indeed, STR contains such an assumption; this has already been mentioned in Chapter 1. STR assumes that the set S is metrically unbounded in the time dimension; both these claims are formulated within the scope of our initial assumptions from STR (cf. Chapter 1). Otherwise, i.e., if, for example, the spatial dimensions of the set S were limited in a metric sense, the paradox of present would hold for certain pairs of events only and not for all pairs.

The following thesis is also important:

$$\text{(T18)} \quad \bigwedge_x \bigwedge_y \bigvee_v \{W(x, y) \rightarrow \sim [x \in N_v \wedge y \in N_v]\}. \qquad \{\text{A5, D2}\}$$

In words: for any two events x and y, there exists such an event v that, if x occurs absolutely earlier than y, then these events do not occur together in the present of the event v. For instance, this is the case when x occurs in the past of v, and y occurs in the future of v. (A5) and (D2) also imply the thesis which differs from (T18) only by the fact that $\breve{W}(x, y)$ is substituted for $W(x, y)$ in its antecedent.

Certain unusual circumstances which illustrate the paradox of present will now be described. Let certain events x and y, satisfying the condition $W(x, y)$, belong to a particular thing a, i.e., $x, y \in a$. To simplify the problem, it is assumed that x is the beginning of a in time, and y is its end in time. All other events belonging to the thing a occur in time (absolutely) between x and y. The thing a—like any other thing—is, according to eventism, the set of all such events, obviously including x and y, i.e., a is identical to its history. Then, according to (T17), there exists such an event z that the whole thing a, i.e., its history, belongs to the present of the event z! This can be formally stated as $a \subset N_z$.

Let us assume that the thing a is the Earth from its beginning to its current state, the human race (i.e., its history until now) included. Then, according to this conclusion, there exists such an event z that the whole Earth, the human race included, belongs to the present of the event z. Since in this case the time interval concerned occupies several billion years, the event z must be very distant (spatially) from the Earth. The distance can be calculated according to formulae of STR.

According to the theorem (T18), there exists such an event v that not the whole thing a belongs to the present of the event v. Applying this conclusion to the example here, a situation is reached in which not the whole of the Earth belongs to the present of v. It may happen that part Z_1 of the Earth belongs to the past of v, part Z_2 belongs to its present, while part Z_3 belongs to its future. We mean here obviously the time-parts of the Earth (and humanity), and we assume that Z_2 lies (absolutely) between Z_1 and Z_3 in time. In this case, as well as in the other cases, the event v need not be very distant from the Earth; it may, for instance, occur on the Sun.

The further theses of PNF theory will now be introduced. First, the sets P_x, N_x, F_x taken separately do not cover the set S, i.e., they are only certain proper parts of this set:

(T19) $\bigwedge_{x} \bigvee_{y'} [y' \notin P_x]$, {T10, T16}

(T20) $\bigwedge_{x} \bigvee_{t'} [t' \notin N_x]$, {T11, T14}

(T21) $\bigwedge_x \bigvee_{z'} [z' \notin F_x]$. {T10, T14}

Second, the self-evident theses below are important:

(T22) $\bigwedge_x \bigwedge_y [y \in P_x \equiv x \in F_y]$, {D3, D5, D1}

(T23) $\bigwedge_x \bigwedge_t [t \in N_x \equiv x \in N_t]$, {D4, T5}

(T24) $\bigwedge_x \bigwedge_z [z \in F_x \equiv x \in P_z]$. {D3, D5, D1}

Third, we have:

(T25) $\bigwedge_x \bigvee_y [x \in P_y]$, {T16, T24}

(T26) $\bigwedge_x \bigvee_t [x \in N_t]$, {T15, T23}

(T27) $\bigwedge_x \bigvee_z [x \in F_z]$. {T14, T22}

These state, respectively, that every event occurs in the past of a certain event (T25); every event occurs in the present of a certain event (T26); every event occurs in the future of a certain event (T27). Finally, we have

(T28) $\bigwedge_x \bigvee_{y'} [x \notin P_{y'}]$, {T21, T24}

(T29) $\bigwedge_x \bigvee_{t'} [x \notin N_{t'}]$, {T20, T23}

(T30) $\bigwedge_x \bigvee_{z'} [x \notin F_{z'}]$. {T19, T22}

These state, respectively, that for every event x such an event y' exists that x does not occur in its past (T28); for every event x such an event t' exists that x does not occur in its present (T29); for every event x such an event z' exists that x does not occur in its future (T30). Let us note that (T25) and (T28) together express the relational character of the past, i.e., the fact that the same events occur in the past of a certain event and do not occur in the past of some other event. The theorems (T26) and (T29) together express the relational character of the present, while the theorems (T27) and (T30) together express the relational character of the future. An abridged version of PNF theory may be found in Augustynek (1976).

The axiomatic system of PNF theory comprises finally the axioms and theses concerning the causal relation H and the connections (based on STR) between this relation and the relation W together with other absolute time relations. We obviously assume that the relation H is defined on the set of all events S. This being so, the formula $H(x, y)$ reads: the event x is a cause of the event y, while the formula $H'(x, y)$ reads: the event x is a result of the event y.

First of all it is assumed that the causal relation H is asymmetric and transitive, and therefore irreflexive in S:

(A11) $H \in \mathbf{asymm}$,

(A12) $H \in \mathbf{trans}$,

(T31) $H \in \mathbf{irrefl}$. {A11, A12}

Consequently, the relation H partially orders the set S; since it is not connected in S, which is not explicitly stated here, it does not impose a linear ordering in S. The relation H' is defined by means of the relation H according to the mentioned meaning of the expression $H'(x, y)$:

(D6) $H' \overset{\text{def}}{=} \breve{H}$.

The relation H' is therefore a converse of the relation H, so the symbol $\breve{H}$ will be used henceforth to denote it. This relation too is asymmetric and transitive in the set S, so it is irreflexive and partially orders the set S:

(T32) $\breve{H} \in \mathbf{asymm}$, {A11, D6}

(T33) $\breve{H} \in \mathbf{trans}$, {A12, D6}

(T34) $\breve{H} \in \mathbf{irrefl}$. {T31, D6}

Let us note an important fact: according to STR, the relations H and $\breve{H}$—as well as relations W and $\breve{W}$—are absolute, i.e., independent of an inertial reference system. Also, these relations are regarded as representing physical interactions between events.

Other important axioms concerning the causal relation will now be formulated. They are as follows:

(A13) $\bigwedge_x \bigvee_y [H(y, x)]$,

(A14) $\bigwedge_x \bigvee_z [H(x, z)]$.

In words: (A13) states that every event has a certain event as its cause, (A14) states that every event has a certain event as its result. It can be said that, in the language of the theory of relations, these axioms express the following: there is no first element with respect to the relation H in the set S (A13), and there is no last element with respect to the relation H

in S (A14). In other words: the set of events S has neither its causal beginning nor its causal end; or, to put it another way, there is no 'first' cause, nor 'last' result. The axiom (A13) is usually called the *ordinary principle of determinism* or *causality*; this latter name will be used here. However, it will denote both the axioms (A13) and (A14) since they are equally important. It is also worth pointing out that they are logically independent—something that often passes unnoticed. Finally, let us add that they are isomorphic with the theses (T14) and (T15) concerning the relation absolutely earlier, W. The problem of their mutual association will be discussed below.

There are two kinds of axioms and theses concerning the connections between the causal relation H (and $\breve{H}$), and the time relation W (as well as $\breve{W}$ and R): one non-modal, which does not make use of the operator of physical possibility, and one modal which does. Let us begin with the non-modal axioms:

(A15) $\bigwedge_x \bigwedge_y [H(x, y) \rightarrow W(x, y)]$,

(T35) $\bigwedge_x \bigwedge_y [\breve{H}(x, y) \rightarrow \breve{W}(x, y)]$. {A15, D6}

In words: if x is a cause of y, then x is absolutely earlier than y (A15), and if x is a result of y, then x is absolutely later than y (T35). In physics the theorems (A15) and (T35) are called the *causal postulate*. This postulate has been generally accepted in modern physics since the emergence of STR.

The following theses are immediately implied by the theorems:

(T36) $\bigwedge_x \bigwedge_y [\sim W(x, y) \rightarrow \sim H(x, y)]$, {A15}

(T37) $\bigwedge_x \bigwedge_y [\sim \breve{W}(x, y) \rightarrow \sim \breve{H}(x. y)]$, {T35}

(T38) $\bigwedge_x \bigwedge_y [R(x, y) \rightarrow \sim H(x, y) \wedge \sim \breve{H}(x, y)]$. {T36, T37, D2}

The last thesis reads: if x is quasi-simultaneous with y, then x is neither a cause nor a result of y. Since the inclusion $R_u \subset R$ holds, the analogous theorem is true of the simultaneity relation R_u. It is worth noting that the ordinary principle of causality ((A13) and (A14)), and the causal postulate (A15) as well as its consequence (T35) imply, respectively, the theorem about the non-emptiness of the set P_x and the theorem about the non-emptiness of the set F_x (i.e., the theses (T14) and (T15)). This means that if the set S has neither a causal beginning, i.e., a 'first' cause, nor a causal end, i.e., a 'last' result (the ordinary principle of causality), then, on the basis of the causal postulate, the set S possesses neither a temporal beginning nor a temporal end. Thus, if it possesses a temporal beginning or end, then it also possesses a causal beginning or end; this theorem is a reciprocal of the previous implication.

The modal theorems concerning these connections will now be introduced. The symbol $\Diamond$ denoted here the operator of physical possibility. As regards its usage, only two conditions are assumed: that physical possibility entails logical possibility, and that $p \rightarrow \Diamond p$. This being the case, it is assumed that $p \rightarrow \Diamond' p$, where $\Diamond'$ denotes the operator of logical possibility. The modal theorems are the following:

(A16) $\bigwedge_x \bigwedge_y [W(x, y) \equiv \Diamond H(x, y)]$,

(T39) $\bigwedge_x \bigwedge_y [\breve{W}(x, y) \equiv \Diamond \breve{H}(x, y)]$, {A16, D1, D6}

(T40) $\bigwedge_x \bigwedge_y [R(x, y) \equiv \sim \Diamond H(x, y) \wedge \sim \Diamond \breve{H}(x, y)]$. {A16, T39, D2}

In words: x is absolutely earlier than y iff x can be a cause of y (A16); x is absolutely later than y iff x can be a result of y (T39); x is quasi-simultaneous with y iff x cannot be the cause of y and x cannot be the result of y, i.e., they cannot be causally connected (T40). Thus, if two events stand in a quasi-simultaneity relation, then they are not only not causally connected (T38), but also cannot be (physically!) connected in this way (T40). Let us note that given $R_u \subset R$ we have the following: if x is simultaneous with y, then x can be neither a cause nor a result of y.

These three theorems show that the absolute time relations $W, R, \breve{W}$ are univocally connected with the causal relation H, or more precisely, with the possibility or impossibility of satisfying the relation by pairs of events. When we speak of a univocal connection, we mean the equivalence relationship obtaining in the three theorems. In the case of similar theorems concerning the connection between the system-dependent time relations W_u, R_u, $\breve{W}_u$, and the possibility or impossibility of satisfying the causal relation by pairs of events, this relationship is replaced by implication, and the connection is not univocal. Hence, given $W \subset W_u$ and (A16), we obtain $\Diamond H \subset W_u$, given $\breve{W} \subset \breve{W}_u$ and (T39), we obtain $\Diamond \breve{H} \subset \breve{W}$, and finally, given $R_u \subset R$ and (T40), we obtain $R_u \subset \sim \Diamond H \cap \sim \Diamond \breve{H}$.

The consequence of these relationships are causal and modal definitions of absolute past, present, and future, i.e., of the sets P_x, N_x, F_x:

$$P_x \overset{\text{def}}{=} \{y \in S\colon \Diamond H(y, x)\},$$ {D3, A16}

$$N_x \stackrel{\text{def}}{=} \{y \in \mathbf{S}\colon \sim \Diamond H(y, x) \wedge \sim \Diamond \breve{H}(y, x)\}, \qquad \{\text{D4, T39}\}$$

$$F_x \stackrel{\text{def}}{=} \{y \in \mathbf{S}\colon \Diamond \breve{H}(y, x)\}. \qquad \{\text{D5, T40}\}$$

In words: the past of the event x, i.e., P_x, is the set of all such events y that can be causes of the event x; the future of the event x, i.e., F_x, is the set of all such events y that can be results (results of action) of the event x, finally, the present of the event x, i.e., N_x, is the set of all such events y that can be neither causes nor results of the event x (i.e., cannot interact with it). The above causal and modal definitions of P_x, N_x, F_x express the property of these sets known as 'causal determination'. It is clear that analogous definitions for the sets P_x^u, N_x^u, F_x^u cannot be formulated since they are defined by means of system-dependent time relations. These sets are not therefore causally determined even though, as we know, they have parts (their subsets) which are—namely, the sets P_x, N_x, F_x.

Within the framework of the non-standard PNF theory presented here, notions of absolute past, present, and future may be introduced referring to things treated in an event-like manner. They are constructed analogously to the corresponding notions of $\mathrm{P^uN^uF^u}$ theory. The symbols $P_{a|x}, N_{a|x}, F_{a|x}$ denote the past of the thing a with respect to the event x, the present of the thing a with respect to the event x and the future of the thing a with respect to the event x. These sets are defined as follows:

$$P_{a|x} \stackrel{\text{def}}{=} \{y \in a\colon\ W(y, x)\},$$

$$N_{a|x} \stackrel{\text{def}}{=} \{y \in a\colon\ R(y, x)\},$$

$$F_{a|x} \stackrel{\text{def}}{=} \{y \in a\colon\ \breve{W}(y, x)\}.$$

In words: $P_{a|x}$ is the set of events which belong to the thing a and are absolutely earlier than x; $N_{a|x}$ is the set of events which belong to the thing a and are quasi-simultaneous with x; $F_{a|x}$ is the set of events which belong to the thing a and are absolutely later than the event x. Hence the sets $P_{a|x}, N_{a|x}, F_{a|x}$ consist of only those events which belong to the thing a, and we therefore speak about the past, present, and future of a particular thing.

Unlike $P^u_{a|x}, N^u_{a|x}, F^u_{a|x}$, these sets are absolute, i.e., system-independent, because the relations W, R and $\breve{W}$ which define them are absolute. It should be added here that the following connections between the sets $P_{a|x}, N_{a|x}, F_{a|x}$ and $P^u_{a|x}, N^u_{a|x}, F^u_{a|x}$ occur: $P_{a|x} \subset P^u_{a|x}$ and $F_{a|x} \subset F^u_{a|x}$, but $N^u_{a|x} \subset N_{a|x}$. Let us note that $P_{a|x}, N_{a|x}, F_{a|x}$ are, as can be seen from their definitions, temporal parts of things.

It should be emphasized that within PNF theory it is not possible to introduce and develop notions of past, present, and future referring to moments, something that can be done within $P^uN^uF^u$ theory (cf. Chapter 2). The reason for this is simple: the relation of quasi-simultaneity which appears in PNF theory is not an equivalence relation. In consequence, it is not possible to define moments themselves (by abstraction) and, consequently, to define the past, present, and future related to moments. This is not a deficiency of PNF theory, since I cannot see any serious reason for applying those notions when the well defined sets P_x, N_x, F_x are available within this theory. This implies another important conclusion: the notions discussed above have a strictly relative character and therefore have their *raison d'etre* only within $P^uN^uF^u$ theory, as the sets P_x^u, N_x^u, F_x^u.

A system-independent theory of present and non-present can easily be formulated analogously to the system-dependent one (cf. Chapter 2). Arguments in favour of doing this can be either the refutation of a physical difference between absolute past and absolute future, or disregard of this difference — a theory ignoring this difference could be interesting in itself. Because of its simplicity, this theory is worth constructing independently of PNF theory as is the case of the $\overline{N}^uN^u$ theory. It will be designated for short as the $\overline{N}N$ theory.

The primitive terms of this system are the symbols **S** and **T**, denoting the sets of events and things, and the symbol $\overline{R}$, denoting the time relation called *absolute time separation* or *non-quasi-simultaneity*, which is defined on the set S. Let us note that if we were constructing our theory within PNF theory, the relation $\overline{R}$ would be defined as the sum $W \cup \breve{W}$, i.e., the complement of the quasi-simultaneity relation R.

The relation $\overline{R}$ is irreflexive, symmetric, and neither transitive nor intransitive in the set **S**:

(A1′) $\overline{R} \in \mathbf{irrefl}$,

(A2′) $\overline{R} \in \mathbf{symm}$,

(A3′) $\overline{R} \notin \mathbf{trans}$,

(A4′) $\overline{R} \notin \mathbf{intrans}$.

The absolute relation of quasi-simultaneity is regarded as derivative, in its definition, from the relation $\overline{R}$: it is defined as a complement of $\overline{R}$, i.e., $R^* = \overline{\overline{R}} = R$ (this last symbol will be used hereafter):

(D1′) $R^* \stackrel{\text{def}}{=} R$.

The relation R is reflexive, symmetric and, something that is additionally assumed, neither transitive nor intransitive:

(T1′) $R \in$ **refl**, {D1′, A1′}

(T2′) $R \in$ **symm**, {D1′, A2′}

(A5′) $R \notin$ **trans**,

(A6′) $R \notin$ **intrans**.

Thus the relation R in the set **S** is not a time-equality but only a time-similarity. However, the relation of the absolute time separation $\bar{R}$ constitutes a time-difference. It is obvious that $Id \subset R$, where Id is the relation of logical identity; this being the case, we have $\bar{R} \subset \overline{Id}$, i.e., the time-difference (separation) is included in the logical non-identity. When the relationships between the relative time relations $\bar{R}_u$ and R_u and the absolute time relations $\bar{R}$ and R are examined, it is not difficult to establish what follows. First, $\bar{R}$ is the product of all the relations $\bar{R}_u, \bar{R}_{u_1}, \bar{R}_{u_2}, \ldots$, while R is the sum of all the relations $R_u, R_{u_1}, R_{u_2}, \ldots$ Second, from the above it follows that the connections $\bar{R} \subset \bar{R}_u$ and $R_u \subset R$ hold.

Obviously, the operational definition of the relation R introduced in PNF theory is also accepted here. This definition provides the operational definition of the relation $\bar{R}$ by means of the negation of the sentence $Q(x, y)$ (concerning the meeting of light signals between points). As a consequence, the symbols $\bar{R}$ and R both acquire an operational meaning.

As in the $\bar{\mathrm{N}}^u\mathrm{N}^u$ theory, the non-present and present are defined by means of the absolute relations $\bar{R}$ and R. As a result, these objects are related to particular events, i.e., they have a relational character, and they are absolute, i.e., independent of the inertial system of reference. The symbols $\bar{N}_x$ and N_x are therefore introduced to denote the absolute non-present of the event x, and the absolute present of the event x; however, the term 'absolute' will not be used further in this context. If the $\bar{\mathrm{N}}\mathrm{N}$ theory is treated as a part of PNF theory, then the non-present $\bar{N}_x$ is simply the sum of the sets P_x and F_x.

Here are the required definitions:

(D2′) $\bar{N}_x \overset{\text{def}}{=} \{y \in \mathbf{S}\colon \bar{R}(y, x)\}$,

(D3′) $N_x \overset{\text{def}}{=} \{y \in \mathbf{S}\colon R(y, x)\}$.

In words: the non-present of the event x, i.e., $\bar{N}_x$, is the set of all events y which are absolutely time-separated from (non-quasi-simultaneous with)

the event x; the present of the event x, i.e., N_x is the set of all events y which are quasi-simultaneous with x. The physico-geometrical interpretation of the sets $\overline{N}_x$ and N_x is very simple. Using the previously introduced notation for the interpretation of the sets P_x, N_x, F_x, the following connections can be established: $\overline{N}_x = (\alpha_1 \cup \alpha_2) \cup (\gamma_1 \cup \gamma_2)$, i.e., it covers the upper and the lower internal part of the light cone $C(x)$, together with their respective surfaces (obviously without the event x); $N_x = \beta \cup \{x\}$, i.e., it covers the region outside the light cone and the set $\{x\}$. This interpretation can be simplified by designating the internal part (lower and upper) of the light cone by α and its surface (also upper and lower) by γ; then we have: $\overline{N}_x = \alpha \cup \gamma$ and $N_x = \beta \cup \{x\}$.

Further axioms and theses corresponding to the equivalent axioms and theses of the $\overline{\mathrm{N}}^{\mathrm{u}}\mathrm{N}^{\mathrm{u}}$ theory will now be formulated. First:

(T3′) $\bigwedge_x [x \notin \overline{N}_x]$, {D2′, A1′}

(T4′) $\bigwedge_x [x \in N_x]$. {D3′, T1′}

This means that an event does not occur in its non-present, but does occur in its present. This expresses a very important difference between $\overline{N}_x$ and N_x. Second, the sets $\overline{N}_x$ and N_x are disjoint and together cover the set $\mathbf{S}$:

(T5′) $\bigwedge_x [\overline{N}_x \cap N_x = \emptyset]$, {D2′, D3′, D1′}

(T6′) $\bigwedge_x [\overline{N}_x \cup N_x = \mathrm{S}]$. {D2′, D3′, D1′}

The connections between the sets $\overline{N}_x, N_x$, and the sets $\overline{N}_y, N_y$, where x and y are arbitrary events, are of great importance. These connections are the following:

(A7′) $\bigwedge_x \bigwedge_y [\overline{N}_x \cap \overline{N}_y \neq \emptyset]$,

(A8′) $\bigwedge_x \bigwedge_y [\overline{R}(x, y) \vee R(x, y) \wedge x \neq y \rightarrow \overline{N}_x \cap N_y \neq \emptyset]$,

(A9′) $\bigwedge_x \bigwedge_y [N_x \cap N_y \neq \emptyset]$.

This means that the non-presents of the arbitrary events x and y always have a non-empty product (A7′). The same is true of the presents of these events (A9′). However, the product of the non-present of the event x and the present of the event y is non-empty, provided that x and y are absolutely time-separated (A8′). The most important difference between the $\bar{N}N$ theory and the $\bar{N}^u N^u$ theory is the axiom (A9′) which expresses the non-emptiness of $N_x \cap N_y$ regardless of the time relation between x and y.

(A7′) implies that the non-present of the event x is non-empty, while (A9′) implies that the present of the event x is non-empty:

(T7′) $\bigwedge_{x} [\bar{N}_x \neq \emptyset]$, {A7′}

(T8′) $\bigwedge_{x} [N_x \neq \emptyset]$. {A9′}

Furthermore, the sets $\bar{N}_x$ and N_x separately do not cover the set **S**:

(T9′) $\bigwedge_{x} \bigvee_{y'} [y' \notin \bar{N}_x]$, {T5′, T8′}

(T10′) $\bigwedge_{x} \bigvee_{z'} [z' \in N_x]$. {T5′, T7′}

Secondly, the following theses are important:

(T11′) $\bigwedge_{x} \bigwedge_{y} [y \in \bar{N}_x \equiv x \in \bar{N}_y]$, {D2′, A2′}

(T12′) $\bigwedge_{x} \bigwedge_{y} [y \in N_x \equiv x \in N_y]$. {D3′, T2′}

We also have

(T13′) $\bigwedge_{x} \bigvee_{y} [x \in \bar{N}_y]$, {T7′, T11′}

(T14′) $\bigwedge_{x} \bigvee_{z} [x \in N_z]$. {T8′, T12′}

These state that every event occurs in the non-present of a particular event (T13′), and every event occurs in the present of a particular event (T14′). Finally, we have

(T15′) $\bigwedge_{x} \bigvee_{y'} [x \notin \bar{N}_{y'}]$, {T9′, T11′}

(T16′) $\bigwedge_{x} \bigvee_{z'} [x \notin N_{z'}]$. {T10′, T12′}

These state that every event does not occur in the non-present of a particular event (T15′); every event does not occur in the present of a particular event (T16′).

The theses (T13′) and (T15′) together express the relational character of the non-present, i.e., the fact that the same events occur in the non-present of a particular event and do not occur in the non-present of some other event. The theses (T14′) and (T16′) together express the relational character of the present—a property analogous to that described above. A more developed outline of the $\bar{N}N$ theory is contained in Augustynek (1982).

Within the $\bar{N}N$ theory one can also introduce the absolute notions of non-present and present referring to things. The non-present of the thing a with respect to the event x can be defined by means of the relation $\bar{R}$, while the relation R can be used to define the present of the thing a with respect to the event x. Being the temporal parts of things, these sets defined thus are obviously non-empty and disjoint, and their sum covers the thing a.

However, the absolute notions of non-present and present related to moments cannot be introduced within the $\bar{N}N$ theory, in contrast to the $\bar{N}^uN^u$ theory, where this is possible. The quasi-simultaneity relation R cannot be used to define moments, so that it is not possible to define either the non-present or the present related to moments. This is an analogous situation to that described above for PNF theory.

Let us now consider the absolute space relations analogous to the absolute time relations $\bar{R}$ and R. Earlier we mentioned the following system-dependent space relations: the relative space separation of events $\bar{L}_u$ and the relative space co-location of events L_u. We asserted that these are sets of concrete relations corresponding to the respective inertial reference systems. We also concluded that $\bar{L}_u$ is irreflexive and symmetric (but neither transitive nor intransitive), i.e., it is a relation of relative space-difference, while the relation L_u is reflexive, symmetric and transitive, i.e., a relation of relative space equality (cf. Chapter 2).

Apart from these relative space relations, STR contains absolute, system-independent space relations also defined on the set S. The first is the absolute relation of space separation $\bar{L}$ and the second the absolute relation of co-location which will be called the *relation of quasi-co-location*, and denoted by $\bar{L}$. Let us note that so far the operational definitions of the relations $\bar{L}$ and L (like those of $\bar{L}_u$ and L_u) are not known.

The initial relation $\bar{L}$ is in STR assumed to be irreflexive and symmetric in the set S and, apart from this, neither transitive nor intransitive; in other words, it is a relation of absolute space difference in S, and consequently analogous to the relation $\bar{R}$ of absolute time difference.

The second relation—that of quasi-co-location L—is defined here, as is clear from the notation applied, as the complement of the relation $\bar{L}$ (the reverse procedure is also possible: the definition of $\bar{L}$ by means of L). This definition implies that the relation L is reflexive and symmetric in the set S, and it is additionally postulated that it is neither transitive nor intransitive in this set. The quasi-co-location L is therefore only a relation of space similarity in S and not, like L_u, a relation of space equality. Nevertheless, in view of the properties mentioned above, it is similar to the relation of quasi-simultaneity R.

It is not out of place here to make the following comment. In the case of absolute time relations, the absolute relations W and $\breve{W}$ can be the starting point since they allow one to define the relations $\bar{R}$ and R, as $W \cup \breve{W}$ and $\overline{W} \cap \overline{\breve{W}}$. The question arises whether, in the case of absolute space relations, the starting point can be some spatial (and absolute) analogues of these time relations W and $\breve{W}$. There are, certainly, such relations as 'to the left of' and 'to the right of'; whether they are absolute requires some explanation which will not be gone into here. These relations can, however, be introduced in this context, and the definitions of absolute space separation $\bar{L}$ and quasi-co-location L can be based on them. This is a procedure similar to that of defining the absolute time relations $\bar{R}$ and R by means of the relations W and $\breve{W}$. The point is, however, that space is three-dimensional and the number of such initial space relations by means of which the other space relations are defined is six, not two. This would imply, in this context, unnecessary complications which we prefer to avoid, especially as we are concerned here only with the spatial analogues of the absolute temporal relations $\bar{R}$ and R.

The connections between the relations $\bar{L}$ and $\bar{L}_u$ and L and L_u given below allow a better understanding of the absolute relations $\bar{L}$ and L. The following two theorems can be proved on the basis of STR. First, the relation $\bar{L}$ is the product of all relations $\bar{L}_u, \bar{L}_{u_1}, \bar{L}_{u_2}, \ldots$, characteristic of the corresponding inertial reference systems. Secondly, this fact and the definitions of the relations L and L_u (as complements of $\bar{L}$ and $\bar{L}_u$) imply that the relation L is the sum of all relations $L_u, L_{u_1}, L_{u_2}, \ldots$, characteristic of the corresponding inertial reference systems. The first theorem implies that the relation of absolute space separation $\bar{L}$ is a (proper) subrelation of the relation of relative space separation $\bar{L}_u$. The second theorem implies that the relation of co-location L_u is a (proper) subrelation of the relation of quasi-co-location $\bar{L}$. Strictly speaking, every relation L_u is a particular subrelation of the relation L. It is clear that the connections between $\bar{L}$ and $\bar{L}_u$, as well as between L and L_u, are fully isomorphic with the connections which hold for the pairs of time relations $\bar{R}, \bar{R}_u$, and R, R_u.

A comparative analysis will now be carried out of the pairs of absolute time and space relations which are similar with respect to their structural properties, i.e., R and L, as well as $\bar{R}$ and $\bar{L}$.

As far as the first pair is concerned, these relations are of similarity only (reflexive, symmetric, but not transitive) as was seen earlier; the first relation, quasi-simultaneity, R, is one of time similarity and the second, quasi-colocation, L, is one of space similarity. Both include, as we know, the corresponding congruence relations, i.e., equalities (reflexive, symmetric and transitive). R contains the time congruence R_u, and L the space congruence L_u; R_u and L_u are, however, system-dependent relations. Let us note that both these relations are implied by the relation of logical identity Id. This means that $Id \subset R$, and $Id \subset L$, which follows from the fact that we have $R_u \subset R$ and $Id \subset R_u$, as well as $L_u \subset L$ and $Id \subset L_u$. In words: if two events are (logically) identical, then they are temporally and spatially similar. This fact will be useful when it comes to considering some problems later on.

As far as the second pair of relations $\bar{R}$ and $\bar{L}$ is concerned, then, as we also saw earlier, they turn out to be relations of difference (irreflexive, symmetric, and non-transitive). The first, absolute time separation, $\bar{R}$, is a relation of time difference, while the second, absolute space separation, $\bar{L}$, is a relation of space difference. Both are, as we know, subrelations of the corresponding system-dependent relations of difference (which have the same properties): $\bar{R}$ is a subrelation of the time relation $\bar{R}_u$ and $\bar{L}$ of the space relation $\bar{L}_u$. It should be emphasized that both relations imply the relation of logical difference $\overline{Id}$. This means that $\bar{R} \subset \overline{Id}$, and $\bar{L} \subset \overline{Id}$ since $Id \subset R$, and $Id \subset L$. In words: if two events are time- or space-different, then they are also logically different.

Absolute space-time relations deserve some attention at this point. In the context of the previous considerations, they are products of compatible (i.e., not mutually exclusive) pairs of relations from the basic relations $\bar{R}$, R, $\bar{L}$, L. There can be (and actually are) four such relations: $R \cap L$, $R \cap \bar{L}$, $\bar{R} \cap L$, $\bar{R} \cap \bar{L}$; these relations are obviously mutually exclusive. It will readily be seen that any two events can stand in only one of these relations. The relations therefore cover the range of all absolute space-time relations (of the kind made out of products of the initial relations). Let us also note that this situation implies that none of the initial relations includes any other one. This conclusion concerns, first of all, the pairs R and L, R and $\bar{L}$, $\bar{R}$ and L, $\bar{R}$ and $\bar{L}$. The pairs R and $\bar{R}$, as well as L and $\bar{L}$, never stand in the relation of inclusion for logical reasons, since their constituents exclude each other.

Something should be said about certain structural properties of these space-time relations. The relation $R \cap L$ is reflexive and symmetric, which follows from the properties of R and L. The relation $\overline{R} \cap \overline{L}$ is irreflexive and symmetric, which follows from the properties of $\overline{R}$ and $\overline{L}$. Other properties of these relations should be examined too as should the properties of the relations $R \cap \overline{L}$ and $\overline{R} \cap L$ if this turns out to be necessary.

The relation $R \cap L$ determines the space-time similarity of events which stand in this relation. It holds, provided that the events are identical (the relation is reflexive). It also holds in cases when two events coincide, i.e., when they overlap in space-time. The problem of the relationship between the relation $R \cap L$ and the relation of coincidence will be discussed later. The intermediate 'mixed' relations $R \cap \overline{L}$ and $\overline{R} \cap L$ are space-without-time separation and time-without-space separation; these occur frequently. Finally, the relation $\overline{R} \cap \overline{L}$ is a space-time separation of events. It holds, for example, when the events are different states of a photon. Generally—as we shall see—an event stands in this relation with another only if the former lies on the surface of the light cone of the latter.

In order to understand more clearly which events stand in these space-time relations, let us consider the so-called *cone regions*, which are defined by means of the constituent relations R, $\overline{R}$, L, $\overline{L}$. These regions are related to a given event, and are the sets of events which stand in a given relation to this event. They are defined by the formula $X_x = \{y \in \mathbf{S}: Q(y, x)\}$, where x is the given event, Q the given relation, and X_x a specific region related to x. Let us recall that $C(x)$ is the light cone of the event x, α the internal (upper and lower) part of the cone, γ the (upper and lower) surface of the cone, $\{x\}$ the set comprising x only, and β its external region. The relation of quasi-simultaneity determines—as we know—the region N_x called the (absolute) present of x, i.e., $N_x = \{y \in \mathbf{S}: R(y, x)\}$, while absolute time separation determines the region called the non-present of x, i.e., $\overline{N}_x = \{y \in \mathbf{S}: \overline{R}(y, x)\}$. The relation of quasi-co-location determines the region which may be denoted by V_x so that $V_x = \{y \in \mathbf{S}: L(y, x)\}$, while absolute space separation determines in turn the region which may be denoted by $\overline{V}_x$ so that $\overline{V}_x = \{y \in \mathbf{S}: \overline{L}(y, x)\}$. It is clear that V_x and $\overline{V}_x$ are analogues of N_x and $\overline{N}_x$.

The connections between the cone regions $N_x, \overline{N}_x, V_x, \overline{V}_x$ and the sets

α, γ, $\{x\}$ (which are parts of the cone $C(x)$) as well as β (which is its external region) can easily be established:

(1) $N_x = \{x\} \cup \beta$,

(2) $\bar{N}_x = \alpha \cup \gamma$,

(3) $V_x = \{x\} \cup \alpha$,

(4) $\bar{V}_x = \beta \cup \gamma$.

These connections immediately yield certain implications concerning the partial overlapping of some of the regions defined above:

(1′) $N_x \cap V_x = \{x\}$,

(2′) $N_x \cap \bar{V}_x = \beta$,

(3′) $\bar{N}_x \cap V_x = \alpha$,

(4′) $\bar{N}_x \cap \bar{V}_x = \gamma$.

Let us note that the products $N_x \cap \bar{N}_x$ and $V_x \cap \bar{V}_x$ are obviously empty. However, the remaining products are not empty. Two of them, $N_x \cap V_x$ and $\bar{N}_x \cap \bar{V}_x$, are identical with the 'small' regions, i.e., with $\{x\}$ and γ, respectively. The remaining two, i.e., $\bar{N}_x \cap V_x$ and $N_x \cap \bar{V}_x$, are identical with the 'large' regions, i.e., with α and β, respectively. Therefore, firstly, the non-present of the events x, which is determined by the time separation $\bar{R}$, overlaps to a large extent the region determined by the quasi-co-location L. This indicates a kinship between the temporal and spatial relations time $\bar{R}$ and space L. Secondly, the present of the event x, which is determined by the quasi-simultaneity R, overlaps to a large extent the region determined by the absolute space separation $\bar{L}$. This, in turn, indicates a kinship between the temporal and spatial relations R and $\bar{L}$. These kinships are not trivial and are—in all likelihood—manifestations of the connections between time and space described in STR.

To round off this analysis of space-time relations, let us consider a certain question concerning the relation $R \cap L$. From STR (cf. Chapter 2) it follows that the relation which is the product of the system-dependent relations of simultaneity R_u and co-location L_u, i.e., $R_u \cap L_u$, is an absolute relation independent of an inertial reference system. This relation is usually called the *space-time coincidence* or the *space-time overlapping of events*. I denote it by K, therefore $K = R_u \cap L_u$. The relation K is an equivalence, which follows from its definition and the fact that the constituent relations R_u and L_u are also equivalences. The relation $R \cap L$ is denoted by K', i.e.,

$K' = R \cap L$, to differentiate it from the relation K. Like K, it is, obviously, an absolute relation which follows from the fact that the relations R and L are, at least in my view, absolute. The relation K' has already been called *space-time similarity* because R and L are similarity relations: that is why K' is reflexive and symmetric. Let us note that the question whether the relation K' is an equivalence, i.e., whether it is also transitive, cannot be solved on the ground of the definition of K', since the relations R and L are assumed not to be transitive. This, however, does not imply either that K' is transitive or that it is intransitive.

Here we come to an essential point: namely, what is the connection between the relations K and K'? Earlier it was shown that the following connections hold: $R_u \subset R$ and $L_u \subset L$. Therefore, we have $R_u \cap L_u \subset R \cap L$, i.e., according to the definitions of the relations K and K', we obtain $K \subset K'$. Thus, the space-time coincidence is contained in the space–time similarity. Does not the inverse inclusion $K' \subset K$ also hold so that the formula $K' \subset K$ provides, in consequence, the equality $K' = K$? When we consider the light cone of a certain event x, and events which stand in the relations R and L to the event x, we see that the connection $K' \subset K$ must hold; thus $K' = K$. In other words, we can never have a situation in which $K'(x, y)$ and $\sim K(x, y)$ occur at the same time. However, this is not a proof of the inverse inclusion, i.e., its derivation from the accepted axioms and definitions of a possible theory of time and space relations.

In Chapter 1, where eventism, i.e., the ontology assumed in this work, was discussed, it was assumed that the equivalence definition of a thing is not necessary to maintain this standpoint. It is enough to accept the existence of things (which constitute the set **T**) on a par with the existence of events (which constitute the sets **S**). Things should be treated as certain sets of events (subsets of the set **S**), and certain theorems concerning them, their relationship to events, their overall temporal, spatial, and causal structure, should be introduced. At this point, we will consider for a moment the temporal and spatial structure of things.

I believe that the objects we commonly refer to as 'things' (including, obviously, persons, i.e., living and conscious objects) and consider in temporal and spatial terms require for their characteristics absolute time and space relations. The reasons in support of this opinion will be presented later. It would seem that (1) if something is a thing, then at least two events belonging to it are related by the absolute time separation, i.e., things are absolutely temporally extended; furthermore, (2) if something is a thing, then at least two events belonging to it are related by the absolute space separation, i.e., things are absolutely spatially extended. These theorems are formulated as follows:

(1) $\bigwedge_{a} \bigvee_{x,y} [a \in \mathbf{T} \rightarrow x, y \in a \wedge \overline{R}(x, y)]$,

(2) $\bigwedge_{a} \bigvee_{v,z} [a \in \mathbf{T} \rightarrow v, z \in a \wedge \overline{L}(v, z)]$.

Three comments: Firstly, the assumption that things (elements of the set **T**) are certain sets of events (subsets of the set **S**) implies the identification of the relation of the occurrence of an event within the thing with the set-theoretical relation of the belonging of events to things. This manifests itself in the expressions $y, z \in a$ and $v, z \in a$ which appear in the theorems (1) and (2). Secondly, from an empirical point of view, the time and space extension of things cannot be other than absolute: a given thing cannot be time extended in one reference system and time point-like, i.e., non-extended, in some other one; the same applies to the space extension of things. The system-dependent relations of time and space separation $\overline{R}_u$ and $\overline{L}_u$ cannot therefore stand for the absolute relations R and L in the theorems (1) and (2). This implies the need to use the latter relations to describe the temporal and spatial structure of things. Finally, one can ask: which theory the theorems (1) and (2) should be included in? My answer would be, in the elementary theory of eventism, though this will not be developed here. A theory of this type, should presumably include a part dealing with absolute time and space relations.

*

Now is the right moment to reflect in general on both theories of past, present, and future that have been presented: the standard (system-dependent) and non-standard (system-independent) theory. What concerns us here are the differences, similarities, and connections between these theories, and also whether one should be preferred to the other. The implications for this latter question, which will appear during the analysis of connections between the notions of past, present, and future, and those of time, existence and becoming, will be disregarded for the time being; they will be considered in subsequent chapters.

When I presented the non-standard theory, which I worked out before the standard one, I treated it as an axiom system fully autonomous with respect to the standard system. The point was to prove that such a system, interpreted, of course, could be built within the framework of STR. Moreover, its conceptual network makes possible the unconstrained consideration of the junction of past, present, and future with time, existence, and becoming. The system also turned out to provide a new idea for a relational definition of time. I have already indicated the connections which evidently hold between the basic relations of the non-standard theory: W, R, $\breve{W}$, and those of the standard theory: W_u, R_u, $\breve{W}_u$. Their consequences, namely the connections between the sets P_x, N_x, F_x of the first theory and the corresponding sets P_x^u, N_x^u, F_x^u of the second, have also been indicated. However, these connections were mentioned for the sole purpose of making it easier for the reader who is already familiar with the standard notions to understand the non-standard, absolute notions.

The theories discussed here differ in many respects. The basic series of time relations differ semantically and formally (the structural properties of the relations R_u and R are different). As a result, the series of fundamental notions—of past, present, and future—differ in an analogous fashion; this is particularly true for the notions of present N_x^u and of quasi-present N_x. The basic difference should not, of course, be forgotten here, namely, that the relations and sets of the standard theory are system-dependent, while those of the non-standard theory are absolute. Although the great majority of theorems of the non-standard theory have their isomorphic counterparts in the standard theory, there are, nonetheless, important exceptions. For example, the standard theory contains the axiom (A9): $R_u(x, y) \equiv N_x^u \cap N_y^u \neq \emptyset$, while the non-standard theory contains the axiom (A10): $N_x \cap N_y \neq \emptyset$. This is, by the way, the essential difference between these two systems, which stems from the difference between the relation R and the relation R_u.

Nevertheless, PNF theory can be regarded as an essential part of $P^uN^uF^u$ theory. As a matter of fact, the connections between the absolute relations W, R, $\breve{W}$ and system-dependent relations W_u, R_u, $\breve{W}_u$, and, in consequence, between the absolute sets P_x, N_x, F_x and system-dependent sets P_x^u, N_x^u, F_x^u, are exceptionally strong. For the above series of relations, they can be formulated more precisely (the symbols $\bigcap$ and $\bigcup$ denote the intersection and the union of sets or relations, respectively):

(1) $W = \bigcap_{u \in U} W_u$,

(2) $\breve{W} = \bigcap_{u \in U} \breve{W}_u$,

(3) $R = \bigcup_{u \in U} R_u$.

For the above series of sets we have:

$$(4)\qquad P_x = \bigcap_{u \in U} P_x^u,$$

$$(5)\qquad F_x = \bigcap_{u \in U} F_x^u,$$

$$(6)\qquad N_x = \bigcup_{u \in U} N_x^u.$$

It should be noted that the theorems (4), (5), and (6) follow from the theorems (1), (2) and (3), together with the definitions of the sets P_x^u, N_x^u, F_x^u (by means of the relations W_u, R_u, $\breve{W}_u$) and the definitions of the sets P_x, N_x, F_x (by means of the relations W, R, $\breve{W}$).

The theorems (1), (2), and (3) can be regarded as definitions of the absolute time relations W, R, $\breve{W}$ by means of the system-dependent relations W_u, R_u, $\breve{W}_u$. In consequence, the theorems (4), (5) and (6) can be regarded as definitions of the absolute sets P_x, N_x, F_x by means of the system-dependent sets P_x^u, N_x^u, F_x^u. Those definitions belong obviously to $\mathrm{P^uN^uF^u}$ theory. Thus PNF theory becomes a part of $\mathrm{P^uN^uF^u}$ theory and ceases to be autonomous with respect to the latter. It should be emphasized that the reverse procedure, i.e., the derivation of $\mathrm{P^uN^uF^u}$ theory as a part of PNF theory, cannot be carried out. This latter is conceptually weaker; as we know, it does not contain the notion of the set U of inertial systems. Most of the notions of $\mathrm{P^uN^uF^u}$ theory referring to such systems cannot therefore be expressed within the apparatus of PNF theory.

If PNF theory had been treated as a part of $\mathrm{P^uN^uF^u}$ theory, i.e., if the theorems (1), (2), and (3) had been treated as definitions belonging to the latter, we would have been faced with a uniform theory of past, present, and future. This would contain, of course, both absolute and system-dependent notions of time relations, as well as absolute and relative notions of past, present, and future, and also axioms and theses concerning both groups of notions. This idea is genuinely alluring; in thinking about it while writing these words, I find myself becoming attracted to it.

Let us add here that the systems of non-present and present—the standard $\overline{\mathrm{N}}^u\mathrm{N}^u$ theory and the non-standard $\overline{\mathrm{N}}\mathrm{N}$ theory—which have until now been regarded as independent, could easily be included in a uniform theory. The standard theory could be included on the grounds that the known connection between the system-dependent time separation and the relations earlier and later, $\overline{R}_u = W_u \cup \breve{W}_u$, is regarded as a definition of the relation $\overline{R}_u$. In consequence, the association of non-present and past and future, $\overline{N}_x^u = P_x^u \cup F_x^u$, is regarded as a definition of the set $\overline{N}_x^u$. The inclusion of the non-standard $\overline{\mathrm{N}}$N theory in $\mathrm{P^uN^uF^u}$ theory requires two steps. To begin with, it must be included in PNF theory; this can be done by accepting the connection between the absolute time separation $\overline{R}$ and the absolute relations W and $\breve{W}$: $\overline{R} = W \cup \breve{W}$, as a definition of the relation $\overline{R}$ within PNF theory, and also by accepting the connection between the absolute non-present and the absolute past and future, $\overline{N}_x = P_x \cup F_x$, as a definition of the set $\overline{N}_x$ within PNF theory. PNF theory can now be included in $\mathrm{P^uN^uF^u}$ theory in the same way as above, i.e., on the grounds of (1), (2), and (3).

Here it may be asked why we need the notion of time separation, and therefore, also the notions of non-present (regardless of whether they are relative or absolute). In fact, the difference (if it really exists) between the relations earlier and later, and therefore between past and future, is simply disregarded in certain cases. An example is the description given above of the time-extension of things. In this instance, one not only can, but should, make use of the notion of non-simultaneity.

A far more important question which may arise at this point is the following: why are the notion of absolute time relations and the corresponding notions of absolute past, present, and future introduced at all? Is it not enough to limit ourselves to 'pure' $\mathrm{P^uN^uF^u}$ theory not containing PNF theory as a part of it? To say it is, would seem plausible if one takes into account that fact that the special theory of relativity is involved here and that system-dependent relations play an essential role within its framework. From here it is only a step to the view that only system-dependent notions actually function in this theory.

However, this is a mistaken point of view. Absolute time magnitudes and relations, as well as the sets generated by them, appear and function within STR at least on a par with their relative counterparts. Moreover, there are good reasons for supposing that this group of notions is the fundamental one. It is not that these are the primitive notions of the theory—this distinction belong, as we know from previous considerations, to relative notions. As a matter of fact, within relativistic physics, and indeed in the whole of contemporary physics, greater theoretical importance is given to objects, relations, functions, and magnitudes which are absolute, i.e., independent of an inertial system of reference. The complex and extensive problem of justifying this view will not be undertaken here. We recall only the reason for the priority attached to the absolute relations W, R, and $\breve{W}$, which are the basis of PNF theory. The point is that these relations are uniquely connected with the causal relation H (also an absolute one), while the connection between the system-dependent time relations W_u, R_u, $\breve{W}_u$ and the causal relation does not have such a character. Forms of these connections were extensively discussed earlier and were formulated in appropriate theorems. In consequence, as was also described in detail, the absolute sets P_x, N_x, F_x are, as was previously stated, 'causally determined'; in the case of the relative sets P_x^u, N_x^u, F_x^u, there is no analogous situation.

This circumstance is very important—the causal relation is absolute, and equivalent to a physical interaction between events (interaction, in a broad sense of the word). A certain general reason for the priority of absolute notions in STR is worth indicating; namely, in the next step in the development of relativistic physics, the general theory of relativity (GTR), attempts are made to discard the notion of an inertial system of reference. The reason for this is quite another matter.

CHAPTER 4

PAST, PRESENT, FUTURE AND TIME

A problem which deserves careful attention is that of the relationship between past, present, future and time. In particular, we want to know whether these objects are parts of time (as is usually suggested in the philosophy of time), or perhaps, whether they are not such parts. Clearly, the answer to this question, or rather its truth, depends on three things. Firstly, on our definitions of past, present, and future. Secondly, on our definition of time. Thirdly, it obviously depends on how we determine the relation 'being a part of' in this context. For this reason, I shall begin by dealing with these questions, especially with the last two.

First of all, however, a few words about a general question. The problem formulated above will be considered solely within the framework of the standard and non-standard theories of past, present, and future presented in this work. It makes no difference in this instance whether they are treated separately, or whether the second theory is treated as a part of the first. From this qualification it follows that we will make use only of those definitions of past, present, and future previously introduced, and only of such definitions of time as can be formulated and accepted within these theories. Finally, when we try to determine the relation 'being a part of', we will also do this within the conceptual apparatus of those theories (that of set theory) taking into consideration the acquired definitions of past, present, future, and time.

As we know, within the standard system of $P^uN^uF^u$ theory we deal with three sequences of definitions of the sets P^u, N^u, F^u, more or less differing from each other. We consider them each in turn and recall their formulations. The first of them comprises definitions of these sets related to events, and relativized to inertial reference systems:

$$P_x^u \overset{\text{def}}{=} \{y \in \mathbf{S}\colon\ W_u(y, x)\},$$

$$N^u_x \overset{\text{def}}{=} \{y \in \mathbf{S}: R_u(y, x)\},$$

$$F^u_x \overset{\text{def}}{=} \{y \in \mathbf{S}: \breve{W}_u(y, x)\}.$$

P^u_x, N^u_x, F^u_x are therefore relative sets of events, since they are defined by means of the relative time relations $W_u, R_u, \breve{W}_u$ determined in the set of events $\mathbf{S}$.

The second series consists of the definitions of these sets related to moments, thus also relativized to inertial reference systems:

$$P^u_m \overset{\text{def}}{=} \{n \in \mathbf{S}|R_u: W'_u(n, m)\},$$

$$N^u_m \overset{\text{def}}{=} \{n \in \mathbf{S}|R_u: R'_u(n, m)\},$$

$$F^u_m \overset{\text{def}}{=} \{n \in \mathbf{S}|R_u: \breve{W}'_u(n, m)\}.$$

P^u_m, N^u_m, F^u_m are therefore relative sets of moments, since they are defined by means of the corresponding relative time relations $W'_u, R'_u, \breve{W}'_u$ determined in the set of moments $\mathbf{S}|R_u$.

Finally, the third series consists of definitions of these sets related to things, and relativized to inertial reference systems:

$$P^u_{a|x} \overset{\text{def}}{=} \{y \in a: W_u(y, x)\},$$

$$N^u_{a|x} \overset{\text{def}}{=} \{y \in a: R_u(y, x)\},$$

$$F^u_{a|x} \overset{\text{def}}{=} \{y \in a: \breve{W}_u(y, x)\}.$$

$P^u_{a|x}, N^u_{a|x}, F^u_{a|x}$ are, as one can see, relative sets of events belonging to a given thing, since they are defined in each case by means of the corresponding relative time relations $W_u, R_u, \breve{W}_u$ determined in the set of events. Thus they are the past, present, and future of a given thing.

To recapitulate: we have in $\mathrm{P^uN^uF^u}$ theory three types of relative past, present, and future related successively to events, moments, and things. They consist of events, of moments, and of events belonging to a thing.

Within the non-standard system of PNF theory, we deal with only two different series of definitions of past, present, and future. The first consist of definitions of the sets related to events and independent of an inertial reference system:

$$P_x \overset{\text{def}}{=} \{y \in \mathbf{S}: W(y, x)\},$$

$$N_x \overset{\text{def}}{=} \{y \in \mathbf{S}: R(y, x)\},$$

$$F_x \overset{\text{def}}{=} \{y \in \mathbf{S}: \breve{W}(y, x)\}.$$

The sets P_x, N_x, F_x are, of course, absolute sets of events, because they are defined by means of the appropriate absolute time relations, W, R, $\breve{W}$ determined in the set of events S.

The second series consists of definitions of the sets related to things and to specific events, independent of an inertial reference system:

$$P_{a|x} \stackrel{\text{def}}{=} \{y \in a\colon\ W(y, x)\},$$

$$N_{a|x} \stackrel{\text{def}}{=} \{y \in a\colon\ R(y, x)\},$$

$$F_{a|x} \stackrel{\text{def}}{=} \{y \in a\colon\ \breve{W}(y, x)\}.$$

$P_{a|x}, N_{a|x}, F_{a|x}$ are absolute sets of events belonging to a given thing, since they are defined by means of the absolute time relations W, R, $\breve{W}$ determined in the set of events S. Thus, they are the past, present, and future of a given thing.

It is now time to examine the second question; namely, what definitions of time can be formulated and, obviously, accepted in the frameworks of both theories under consideration. To begin with, a general postulate will be introduced concerning the formal character of the definition of time. As will be seen, this postulate obviously imposes some restrictions on possible definitions of time. It states that time is a set-theoretical relational structure $\langle X, Q\rangle$, where X, the extension of this structure (extension of time), is a set, and Q, the characteristic of this structure (characteristic of time), is a relation determined in X, and linearly or partially ordering X. Without substantiating this further, it would appear that time is a certain set ordered, at least partially, by some relation. We will return to this postulate further on.

To begin with, let us formulate the definitions of time within $\mathrm{P^uN^uF^u}$ theory. First, we have a relational definition of relative time, which ascertains the following:

(DI) $C^u \stackrel{\text{def}}{=} \langle \mathrm{S}, W_u\rangle.$

According to (DI), time, to be denoted here by C^u, is a structure, the extension of which is the set of all events S, while its characteristic is the relation relatively earlier W_u, determined in the set S and partially ordering this set. Here we have in mind a partial ordering, since W_u is asymmetric and transitive, but not connected in S. Stated briefly: time C^u is the set S ordered partially by the relation W_u. Thus, (DI) satisfies our postulate introduced above.

Let us also note what follows: (DI) is, strictly speaking, a definition schema from which specific system-related definitions of time may be obtained by substituting for W_u specific relations corresponding to different definite reference systems u_1, u_2, etc. These different definitions determine different system-related times C^{u_1}, C^{u_2}, etc. The relativity of time established by STR, when time is understood in the way expressed by (DI), can be represented by the following state of affairs: $C^{u_1} \neq C^{u_2}$, when $u_1 \neq u_2$.

The second definition is a non-relational definition of relative time by abstraction:

$$\text{(DII)} \quad C^u_m \overset{\text{def}}{=} \langle \mathrm{S}|R_u, W'_u \rangle.$$

According to (DII), time, which will now be denoted by C^u_m, is a structure, the extension of which is the set of all moments $\mathrm{S}|R_u$, while its characteristic is the relation relatively earlier W'_u determined in the set $\mathrm{S}|R_u$, and linearly ordering this set. Elements of the set $\mathrm{S}|R_u$, i.e., moments, are abstraction classes of the (equivalence) relation simultaneously R_u in the set of events S. Moments are, therefore, specific sets of events, and $\mathrm{S}|R_u$ is a family of such sets. The relation relatively earlier W'_u, determined in the set $\mathrm{S}|R_u$, is defined by means of the relation relatively earlier W_u, determined in the set S. The relation W'_u is asymmetric, transitive, and connected in the set $\mathrm{S}|R_u$, and consequently orders this set. By means of W'_u, further relative relations are defined in the set $\mathrm{S}|R_u$: the relation later $\breve{W}'_u$ and the relation simultaneously R'_u; the latter is the same as logical identity. As we know, all three relations serve the purpose of defining the sets P^u_m, N^u_m, F^u_m.

To recapitulate: time C^u_m is a set of abstraction classes of the relation R_u in the set S, ordered linearly by the relation W'_u. The definition (DII), like (DI), is a definition schema from which one obtains particular system-related definitions of time by abstraction, namely by substituting for R_u (and also for W'_u) specific relations corresponding to different specific inertial reference systems. Let us note that, if $R_{u_1} \neq R_{u_2}$ (and accordingly $\mathrm{S}|R_{u_1} \neq \mathrm{S}|R_{u_2}$), then also $W'_{u_1} \neq W'_{u_2}$, because they are determined in different sets. These various particular definitions determine various system-related times, defined by abstraction. The relativity of time, when time is understood as in (DII), can be expressed by the following situation: $C^{u_1}_m \neq C^{u_2}_m$, when $u_1 \neq u_2$. The STR thesis concerning the relativity of time is usually formulated in this way, provided that one implicitly or explicitly assumes a definition of time by abstraction.

Finally, the third definition is a relational definition of the relative time of things. This has not so far been considered. There would, nevertheless, seem to be good reasons for speaking of the time (relative or absolute) of a particular thing, analogously to the manner in which one speaks of the past, present, and future of a given thing. We propose to formulate this definition as follows:

$$\text{(DIII)} \quad C_a^u \overset{\text{def}}{=} \langle a, W_u \rangle.$$

According to (DIII), the time of a thing a in the system u, which will be denoted here by C_a^u, is a structure, the extension of which is a set-thing a, while its characteristic is the relation relatively earlier W_u determined in the set a and partially ordering this set.

Things are treated here in an event-like manner as specific sets of events, therefore a is such a set. As regards the relation W_u, we assume that it is asymmetric and transitive, but not connected in the set a; consequently, it orders the set–thing a only partially. We also assume that every thing is dense with respect to the relation W_u.

To recapitulate: time C_a^u is the set–thing a, partially ordered by the relation relatively earlier W_u. And again, (DIII) is a definition schema from which one may obtain particular system-related definitions of the time of the thing a, by substituting for W_u specific relations corresponding to different inertial reference systems. These different definitions determine the different system-related times of a given thing. The relativity of the time of the thing a consists in the fact that $C_a^{u_1} \neq C_a^{u_2}$, when $u_1 \neq u_2$. (DIII) is obviously a definition schema in another respect too: we obtain from it particular definition schemata by substituting specific things for a. These different definition schemata determine the different times of different things. For example, if $a_1 \neq a_2$, then $C_{a_1}^u \neq C_{a_2}^u$. We speak here about definition schemata, because each of them comprises different system-related times of corresponding things. The definition is completely determined only if it is clear what thing and which system are involved.

Two comments are necessary at this point. Firstly, time as described above should not be simply identified with the so-called *proper time* of a material particle which is used in STR. The latter is the time of a particle in a system in which it rests. If the thing a rests in the system u, then time C_a^u, specified in the above manner, is the proper time of the thing a in the STR sense. It is one of the times of the thing a, therefore, distinguished from the others, nevertheless a also has other times. Secondly, it should be emphasized that, strictly speaking, (DIII) is not a definition of time, but only a description of the time of an arbitrary given thing which is a part of the universe of events S. On the other hand, the definitions (DI) and (DII) are definitions of time in a strict sense, since they do not include any restrictions concerning the extension of time as a relational structure. (DI) refers to the set of all events S, while (DII) refers to the set of all moments $S|R_u$, elements of which cover the set S.

There are two definitions of time within PNF theory, or, more accurately, only one, because the second is analogous to (DIII) above. These two definitions, which are of some interest, have not been considered so far. The first one will be called a *relational definition of absolute time*. It is as follows:

(DI') $C \stackrel{\text{def}}{=} \langle S, W \rangle$.

According to (DI'), time, denoted here by C, is identical with the structure the extension of which is the set S, and the characteristic of which is the relation absolutely earlier W defined in the set S, and ordering it partially. This last feature stems from the fact that W is asymmetric and transitive in S, but its connectedness is not assumed. More concisely, time C is the set S partially ordered by the relation W. Thus, the definition fulfils our formal postulate.

Let us note that the above definition, in contrast to (DI), (DII), and (DIII) formulated in $P^u N^u F^u$ theory, is not a schema. This results from the fact that the relation W (as well as $\breve{W}$ and R) is absolute, i.e., independent of an inertial reference system. This character of the definition implies that time defined thus, i.e., C, is also absolute and independent of any system. In other words, there is one collective time for all inertial systems of reference. Or to put it in yet another way, there is extra-systemic (or meta-systemic) time. Let us immediately avert possible misunderstandings—the above statement does not contradict STR. The point is that time is understood here differently from the way it is usually conceived in STR, where one accepts—if this is done at all—the definition by abstraction (DII), or—rather infrequently—the relational definition (DI). Both these definitions, as was stated earlier, assume that time is relative.

The second and last definition is the relational definition of the absolute time of things. It is formulated as follows:

(DII') $C_a \stackrel{\text{def}}{=} \langle a, W \rangle$.

According to (DII'), the time of the thing a, denoted here by C_a, is identical with the structure the extension of which is the set–thing a, and the characteristic of which is the relation absolutely earlier W defined in the set–thing a, and ordering this set partially. As above, things are regarded here as specific sets of events. The relation W is assumed to be asymmetric and transitive in the set S, but not connected. Hence, it orders the set–thing a only partially. Things are also assumed to be dense with respect to the relation W.

To sum up the definition concisely: time C_a is the set–thing a partially ordered by the relation absolutely earlier W. The definition is a definition schema in the sense that specific definitions of the absolute time of a thing can be obtained from it by substituting concrete things for a. These different definitions determine various absolute times of different things. For example, if $a_1 \neq a_2$, then $C_{a_1} \neq C_{a_2}$. A definition of this kind is completely specified in cases where a concrete thing is fixed. (DII′) is not a definition schema in the sense related to a reference system. For any particular set–thing a, there is only one relational definition of absolute time. Thus there exists the relational absolute time of the thing a, i.e., one extra-systemic proper time of that thing. This follows, of course, from the fact that the relation earlier W is absolute and, therefore, one relation. There can be many different relative times of a given thing (DIII), but only one absolute time.

The relationship between the absolute time of a thing and the proper time of a particle used in STR requires further examination. At the moment we cannot provide an answer to this problem. Further, what was said as regards (DIII) is also the case for (DII′); strictly speaking, (DII′) is not a definition of time, but only a description of the absolute time of concrete things which are certain parts of the set of events S. On the other hand, like (DI) and (DII), (DI′) is a definition of time in a strict sense, because it does not contain any qualifications concerning the extension of time as a relational structure; (DI) covers the set of all events.

Disregarding for a while the problem dealt with in the present chapter, it is worth comparing the above definitions of time formulated within $P^uN^uF^u_t$ theory and PNF theory. In this way we can point out and discuss their merits and shortcomings, and reflect upon the question of possible preference for one of them. What we have in mind are, first of all, the *sensu stricto* definitions of time, i.e., (DI), (DII), and (DI′).

In the first place, it is necessary to explain the above tacit assumption regarding the division of the definitions of time (and also the definitions of time of things) into relational and non-relational. The set X, i.e., the extension of time as a structure $\langle X, Q\rangle$, can be either (i) a specific set of temporal objects, e.g. moments, or (ii) a set of elements which are not such objects, e.g. events. The fact of whether (i) or (ii) is the case has non-formal character. Let us also note that it is not important whether those specific temporal objects are (or can be) defined by means of objects that are non-temporal, e.g., whether moments are some sets of events or not.

A non-relational definition of time is one where the extension of time is a set of objects of a first type, while a relational definition of time is one where the extension of time is a set of objects of the second type. This classification of definitions is of course disjoint, complete, and non-empty. The definition of time by abstraction (DII), and also the so-called *Newton's definition of time* (analysed in detail in Augustynek, 1975) are both non-relational. On the other hand, definitions of relative (DI) and absolute time (DI′) are relational, as are, obviously, the definitions of relative (DIII) and absolute (DII′) time of things.

A comment is in order here. The term 'relational' is applied to these definitions because a fundamental component of the description of time as a structure $\langle X, Q \rangle$ is not the set X, but the relation Q (determined in it), since X is here a set of non-specific temporal objects of one kind or another, while Q is always a time relation. This is not true of non-relational definitions where both Q and X have a 'temporal' character. In view of this, it is possible to reduce the structure $\langle X, Q \rangle$ to the extension alone, i.e., to the set X (disregarding its characteristic), and define time, for instance, simply as the set of all moments $S|R_u$, while ignoring its ordering. Such a procedure would make no sense where relational definitions are concerned because when we take the definitions (DI) and (DI′) it implies the absurd consequence that the set of events S itself is time! Hence, one can see how important the role of the characteristic Q is here. Clearly, this procedure of reducing a characteristic in the definition of time is tantamount to a weakening of the formal postulate formulated above.

It is not out of place here to recall that Leibniz was the first to propose (in the above-mentioned sense) definition of time that is relational, although not exactly in the version presented here which makes use of the apparatus of set-theory and the ontology of eventism. Hence, this definition is often referred to as the *Leibniz definition.* Of course, Leibniz, who lived in the days of pre-relativistic physics, could not have known that two such definitions are possible—one for relative and one for absolute time.

With reference to the *sensu stricto* definitions of time (DI), (DII), and (DI′), let us consider briefly how to evaluate their merits and shortcomings. Strictly speaking, we are concerned here with evaluating the relational definitions (DI) and (DI′), i.e., the definitions of relative and absolute time, as compared with the non-relational definition of time by abstraction (DII), since this seems to be the most interesting aspect of the matter.

The definition (DI) of relative time has a few essential shortcomings that have been overlooked so far, apart from the one already stressed several times that, contrary to modern physics, it does not employ the concept of the moment, and thus does not allow for an application of the concept of a time interval. These shortcomings are as follows. First, it is not possible to introduce into time defined this way a unique system of coordinates, or the Frechet metric. Secondly, neither translation nor inversion of time can be defined in this case. Accordingly, it is not possible to formulate here the two principal alternatives concerning the properties of symmetry of time, namely whether time is homogeneous or non-homogeneous, and whether time is isotropic or anisotropic. It can easily be shown that the source of both groups of shortcomings is the fact that the relation relatively earlier W_u is not connected in the set **S**.

It is not difficult to demonstrate that the definition of absolute time (DI′) suffers from exactly the same shortcomings; first we have the general drawback that it does not permit the use of the concept of a moment, and—even more important—we have the two groups of defects mentioned above. The source of these latter is the fact that the relation absolutely earlier, which is the basis of this definition, is also not connected in the set **S**. From this point of view, the above definition of time (DI′) is not in any better position than the definition (DI). On the other hand, the non-relational definition of time by abstraction (DII) clearly does not have any of these shortcomings; neither the general deficiency—it explicitly employs the concept of a moment, and allows for the application of the concept of an interval—nor the specific defects pointed out above; this follows from the fact that the relation earlier W'_u is connected in the set $\mathbf{S}|R_u$.

A basic question arises at this point: do these shortcomings, typical of both the relational definitions of time, force us to reject these definitions as unacceptable within STR? Until recently, I was convinced that such a rejection was necessary and that only the definition of time by abstraction which did not have these shortcomings was acceptable. Now, however, after closer investigation of the relational definitions of time, I have changed my mind. Accordingly, I think that the advantages of these definitions can outweigh the shortcomings and that they deserve a serious treatment.

Let us concentrate on the definition of absolute time, since this is where the greatest advantages are apparent. Firstly and basically, this definition extracts and expresses the absolute aspect of time which undoubtedly appears in STR. It does so on the basis of the relation absolutely earlier W. This aspect is, as a general rule, left out of account or even overlooked; it has never been treated in the same way as here—as a starting point to a definition of time (other than by abstraction). In presentations of STR, and in philosophical reflections concerning this theory, what is generally put forward is the relative aspect of time. Usually this is expressed in a definition of time by abstraction (different system-related times C_m^u), very seldom, if at all, in a relational definition of relative time (different system-related times C^u)—and both definitions are based on the relation relatively earlier W_u. The statement that time is relative, of course, only concerns time as determined by one of these definitions. The question of the connection between the relative and absolute aspects of time will be dealt with below.

Secondly, the idea of applying this definition in the domain of GTR seems to be suggested by the definition itself. As is known, time cannot be defined (non-locally) by abstraction in this domain; the relative relation of simultaneity R_u is not transitive here, or simply does not make sense. Probably it is not possible either to define time by means of a similarity relation in this case (such a definition would be non-relational too). Nor, let us add, can the relational definition of relative time (DI) be applied here non-locally. In this situation, we are left only with the relational definition of absolute time. Whether it can actually be applied in the domain of GTR is an open question which needs examination.

If we accept—taking into consideration the above arguments, and treating PNF theory as a part of $P^uN^uF^u$ theory—both the relational definitions of time, then we have three definitions in this uniform theory, since the definition of time by abstraction cannot be ignored. Is this not too many? What functions would justify the existence of these definitions? To eliminate misunderstandings, let us say at once that those definitions determine somewhat different objects, especially the relational definitions, in opposition to the definitions by abstraction. Thus, they are not definitions of the same object, i.e., time, but they are interconnected: the characteristic relations from (DII) and (DI′), i.e., W_u' and W, are defined by means of the characteristic W_u from (DI). Hence, one can say that they express different aspects of time (in a loose sense of the word); e.g. (DI) the relative aspect, and (DI′) the absolute aspect (these aspects do not exclude each other).

It is now appropriate to examine the concept of 'being a part of' in the context of the problem discussed here. Time, defined one way or another, has always been treated here as an ordered or partially ordered set. What we are concerned with, therefore, are parts of exactly such a set. Accordingly, let $\langle X, Q\rangle$ be a relational structure, where X is a set and Q a non-empty relation determined in X and ordering or partially ordering this set. A structure of this kind $\langle X', Q\rangle$ is called a part of a structure $\langle X, Q\rangle$ iff (i) the set X' is a proper subset of the set X, that is $X' \subsetneqq X$, and (ii) the set X' is ordered or partially ordered by the relation Q—obviously confined to the set X'.

In other words, substructures $\langle X', Q\rangle$ of the structure $\langle X, Q\rangle$ are naturally described as parts of the latter.

The conditions (i) and (ii) which appear in this definition are logically independent of each other. Indeed, a situation is possible where (i) holds and (ii) does not; for instance, in the case of a one-element set which cannot be ordered by the relation Q. It should be stressed at this point that Q is construed as an asymmetric, transitive, and connected (ordering) relation, or as an asymmetric and transitive (partially ordering) relation. A situation is also possible where (ii) holds but (i) does not, for instance, in the trivial case when the set X' is identical with the set X, i.e., $X' = X$. Then, according to the definition, $\langle X', Q\rangle$ is not a part of $\langle X, Q\rangle$.

We can now proceed to the solution of the main problem of the present chapter. The question is a simple one: when considering different pairs consisting of a definition of past, present, and future on the one hand, and a definition of time on the other, it is not difficult to find out whether such a pair satisfies the conditions (i) and (ii) imposed above by the definition of the concept of a part. If it does, then past, present, and future are parts of time; if it does not, then they are not such parts. Let us consider first those pairs which clearly do not comply with our definition of a part.

First of all, the sets P_m^u, N_m^u, F_m^u cannot be, and therefore are not, proper parts of the extension of time, neither as the structure $\langle \mathrm{S}, W_u\rangle$, nor as the structure $\langle \mathrm{S}, W\rangle$. This stems from the fact that they are sets of moments (that is, certain sets of the sets of events), and not sets of events. Moreover, these sets cannot be, and therefore are not, ordered by the relation W_u or by W, i.e., by the characteristics specific to those structures. This is the case because the relations W_u and W are—as we know—only determined in the set S. Thus the conditions (i) and (ii) are not satisfied here. Consequently, the sets P_m^u, N_m^u, F_m^u are neither parts of time C^u, nor parts of time C.

Secondly, the sets P_x^u, N_x^u, F_x^u cannot be, and therefore are not, proper parts of the extension $\mathrm{S}|R_u$ of time as the structure $\langle \mathrm{S}|R_u, W_u'\rangle$, since they are sets of events and not of moments, and as such can only be parts of the set S. Besides, they cannot be, and thus are not, ordered by the relation W_u' (the characteristic of this structure), because the relation W_u' is only determined in the set $\mathrm{S}|R_u$. Hence the conditions (i) and (ii) are not satisfied here either, and therefore the sets P_x^u, N_x^u, F_x^u are not parts of time C_m^u.

Thirdly, the sets P_x, N_x, F_x cannot be, and therefore are not, proper parts of the extension $\mathrm{S}|R_u$ of time as the structure $\langle \mathrm{S}|R_u, W_u'\rangle$, for they are sets of events and not of moments. Moreover, they cannot be, and thus are not, ordered by the relation W_u' (the characteristic of this structure), since this relation is only determined in the set of moments $\mathbf{S}|_u R$. And so, the conditions (i) and (ii) are not satisfied here either; consequently, the sets P_x, N_x, F_x are not parts of time C^u_m.

Let us proceed now to the pairs for which the introduced definition holds (although it does not hold completely in one of the following cases). Firstly, let us consider the pair: the sets P^u_m, N^u_m, F^u_m and the structure $\langle \mathbf{S} | R_u, W'_u \rangle$, i.e., time C^u_m. These sets are, of course, proper parts of the extension $\mathbf{S}|R_u$ of time as this structure; thus we have: P^u_m, N^u_m, $F^u_m \subsetneq \mathbf{S}|R_u$. This being so, the condition (i) is satisfied here. The sets P^u_m and F^u_m, which have more than one element (something that can easily be shown on the basis of the $\mathrm{P^uN^uF^u}$ theory), are ordered by the relation W'_u, the characteristic of the structure $\langle \mathbf{S}|R_u, W'_u \rangle$. Hence, there exist the structures $\langle P^u_m, W'_u \rangle$ and $\langle F^u_m, W'_u \rangle$.They are, respectively, the ordered past of the moment m, and the ordered future of the moment m. In this case, the condition (ii) is satisfied too. Consequently, $\langle P^u_m, W'_u \rangle$ and $\langle F^u_m, W'_u \rangle$ are substructures of the structure $\langle \mathbf{S}|R_u, W'_u \rangle$, and, in precisely this sense, are parts of time C^u_m. But what about the present, i.e., the set N^u_m? Though N^u_m is a proper part of the extension of time, it is nevertheless not a part of time C^u_m; as a one-element set (since $N^u_m = \{m\}$), it cannot be ordered by the relation W'_u (or by any other relation), and, consequently, the postulated structure $\langle N^u_m, W'_u \rangle$ does not exist. At this point I refer to a note made by A. Mostowski in his book on logic (Mostowski, 1948, p. 141).

Secondly, let us consider the sets P^u_x, N^u_x, F^u_x and the structure $\langle \mathbf{S}, W_u \rangle$. These sets are clearly proper parts of the extension of time as this structure; thus we have: $P^u_x, N^u_x, F^u_x \subsetneq \mathbf{S}$. Consequently, the condition (i) is satisfied. The sets P^u_x, N^u_x, F^u_x, being such parts and having more than one element (which can be shown within $\mathrm{P^uN^uF^u}$ theory), are partially ordered by the relation W_u, the characteristic of the structure $\langle \mathbf{S}, W_u \rangle$. Hence, there exist the structures $\langle P^u_x, W_u \rangle$, $\langle N^u_x, W_u \rangle$, and $\langle F^u_x, W_u \rangle$. These are, respectively, the partially ordered past of the event x; the partially ordered present of the event x; and the partially ordered future of the event x. Thus the condition (ii) is also satisfied. Consequently, $\langle P^u_x, W_u \rangle$, $\langle N^u_x, W_u \rangle$, $\langle F^u_x, W_u \rangle$ are substructures of the structure $\langle \mathbf{S}, W_u \rangle$, and, in this sense, they are parts of time C^u. Let us, however, note that the relation W_u is empty in the set N^u_x, i.e., in the present of the event x, because this set consists of mutually simultaneous events only.

Finally, let us consider the sets P_x, N_x, F_x, and the structure $\langle \mathbf{S}, W \rangle$. These sets are, obviously, proper parts of the extension $\mathbf{S}$ of time as this structure; thus we have $P_x, N_x, F_x \subsetneq \mathbf{S}$, and the condition (i) is satisfied. The sets P_x, N_x, F_x, being such parts and having more than one element (which can be proven in PNF theory), are ordered by the relation W, i.e., the characteristic of time $\langle S, W\rangle$. Hence, there exist the structures $\langle P_x, W\rangle$, $\langle N_x, W\rangle$, $\langle F_x, W\rangle$. They are respectively: the partially ordered absolute past of the event x; the partially ordered absolute present (quasi-present) of the event x; and the partially ordered absolute future of the event x. Thus, the condition (ii) is satisfied. Consequently, the structures $\langle P_x, W\rangle$, $\langle N_x, W\rangle$, $\langle F_x, W\rangle$ are substructures of the structure $\langle S, W\rangle$, and in this sense they are parts of time C. Let us add parenthetically that, although it is easy to show within PNF theory the non-emptiness of the relation W in the sets P_x and F_x, the author is nevertheless unable to demonstrate the evident non-emptiness of the relation W in the set N_x.

To recapitulate: The ordered sets P_m^u and F_m^u (N_m^u is here an exception) are parts of time defined by abstraction, i.e., C_m^u; the partially ordered sets P_x^u, N_x^u, F_x^u are parts of relational relative time, i.e., C^u; and the partially ordered sets P_x, N_x, F_x are parts of relational absolute time, i.e., C.

To conclude this chapter, I should like to consider our main problem in a slightly different context, namely: (i) instead of speaking of past, present, and future, we will now speak of non-present and present; (ii) we will weaken the formal assumption concerning the definition of time, and impose on the relation Q of the structure $\langle X, Q\rangle$ only the requirement that it should be determined in X, and, obviously, non-empty there; and (iii) non-present and present will be considered as parts of time provided that they are proper subsets of the set X (an extension of time), and a certain time relation Q is determined on them (a characteristic of time).

Under those modified assumptions, we may consider our problem within the framework of $\overline{\mathrm{N}}^u\mathrm{N}^u$ theory and $\overline{\mathrm{N}}\mathrm{N}$ theory which can be, as we know, construed as fragments of $\mathrm{P}^u\mathrm{N}^u\mathrm{F}^u$ theory. Let us limit ourselves here to the non-present and present related to events only. In the $\overline{\mathrm{N}}^u\mathrm{N}^u$ theory we have the following definition of these objects relativized to inertial reference systems:

$$\overline{N}_x^u \overset{\text{def}}{=} \{y \in \mathbf{S}\colon\ \overline{R}_u(y, x)\},$$

$$N_x^u \overset{\text{def}}{=} \{y \in \mathbf{S}\colon\ R_u(y, x)\}.$$

Thus $\overline{N}_x^u$ and N_x^u are relative sets of events since they are defined by means of the respective relative time relations $\overline{R}_u$ and R_u. On the other hand, in the $\overline{\mathrm{N}}\mathrm{N}$ theory we have the following definitions of the discussed objects independent of inertial reference systems:

$$\overline{N}_x \overset{\text{def}}{=} \{y \in \mathbf{S}\colon\ \overline{R}(y, x)\},$$

$$N_x \stackrel{\text{def}}{=} \{y \in \mathbf{S}: R(y, x)\}.$$

$\bar{N}_x$ and N_x are, therefore, absolute sets of events defined by means of the respective absolute time relations $\bar{R}$ and R.

Let us now introduce definitions of time which can be formulated within the frameworks of both mentioned theories. This can be done since, among other things, our formal postulate has a weaker form, namely (ii). In this way, time can be identified with a set with a certain time relation specified in it which, however, neither orders it, nor orders it partially. In the $\bar{\mathrm{N}}^{\mathrm{u}}\mathrm{N}^{\mathrm{u}}$ theory, one can give the following, perhaps the only one possible, relational definition of the relative time:

$$C'^u \stackrel{\text{def}}{=} \langle \mathbf{S}, \bar{R}_u \rangle.$$

Consequently, time is here a structure, the extension of which is the set S and the characteristic of which is the relation of the relative time-separation $\bar{R}_u$ specified in the set S and non-empty there. More concisely: C'^u is a set S in which the relation $\bar{R}_u$ is specified. Let us note that C'^u is, strictly speaking, a definition schema from which one obtains concrete system-related definitions by substituting for $\bar{R}_u$ concrete relations.

In the $\bar{\mathrm{N}}\mathrm{N}$ theory, one can formulate the following (the only possible) relational definition of the absolute time:

$$C' \stackrel{\text{def}}{=} \langle \mathbf{S}, \bar{R} \rangle.$$

Thus time in this case is a structure, the extension of which is the set S and the characteristic of which is the relation of the absolute time-separation specified in S. In short: C' is the set S in which the relation $\bar{R}$ is specified. This definition is not a definition schema because $\bar{R}$, as absolute, is just one relation.

Let us add two comments. First, these definitions are undoubtedly counter-intuitive. Intuition would suggest that time is a certain set ordered at least partially (in other words, an oriented set), whereas the definitions above specify time as an unordered or unoriented set. Nevertheless, this does not seem to me to be an essential difficulty; if one may, in certain circumstances, ignore the physical (nomological) difference between the relations W_u and $\breve{W}_u$, and the relations W and $\breve{W}$, then one can also accept such concepts of time. Besides, the relations $\bar{R}_u$ and $\bar{R}$ in some sense replace the above-mentioned pairs of relations by virtue of the connections already known to us: $\bar{R}_u = W_u \cup \breve{W}_u$ and $\bar{R} = W \cup \breve{W}$. In other words, these relations characterize (in the given definitions) the time extension of the set S, in which they are defined, not indicating an orientation of this extension. Secondly, in the face of what has been said above, one should refer to C^u and C'^u as relative oriented time and relative unoriented time, respectively, C and C' should be spoken of as absolute oriented time and absolute unoriented time.

Finally, let us consider the relationship between the relative and absolute non-present and present, and the formulated definitions of time. The sets $\overline{N}_x^u$ and N_x^u are, of course, proper parts of the set S, and the relation of the relative time-separation is specified in them too. We have, therefore, the structures $\langle \overline{N}_x^u, \overline{R}_u \rangle$ and $\langle N_x^u, \overline{R}_u \rangle$ which are parts of the relative unoriented time, i.e., the structure $\langle S, \overline{R}_u \rangle$. Let us note, however, that the relation $\overline{R}_u$ is empty in N_x^u. The sets $\overline{N}_x$ and N_x are also proper parts of the set S, and the relation of the absolute time-separation is specified in them. Thus there are the structures $\langle \overline{N}_x, \overline{R} \rangle$ and $\langle N_x, \overline{R} \rangle$ which are parts of the absolute unoriented time, i.e., the structure $\langle S, \overline{R} \rangle$. In this case, the relation $\overline{R}$ is non-empty both in $\overline{N}_x$ (which is obvious) and in N_x.

CHAPTER 5

PAST, PRESENT, FUTURE AND EXISTENCE

The next problem to be dealt with is the relationship between existence and past, present, and future. If we accept as certain that present events exist, does it follow that past and future events (and also things) exist as well? That, at any rate, is the classic formulation of the problem.

The problem will, of course, be considered within the framework of the theories of past, present, and future presented here. The standard theory is discussed first, and subsequently the non-standard theory. The method will be to formulate and investigate various assumptions and hypotheses concerning the relationship between existence and past, present, and future, i.e., to examine different solutions put forward. We shall also attempt to evaluate these assumptions in order to make a choice among them. It follows from this framework that only those definitions of past, present, and future will be used which appear in these theories. It will be assumed, however, that the scope of reference of these notions will be limited to events only (and not to moments, for example). This means that we shall operate only with the sets P_x^u, N_x^u, F_x^u and P_x, N_x, F_x. As we shall see, this emphasizes certain specific aspects of the notion of existence, without—in our view—distorting its fundamental meaning.

As regards this notion, let us at once note what follows. Firstly, the notion of existence is in both theories regarded as primitive on a par with such terms as **S**, **T**, and others. This is assumed not only because it would appear (indeed, as supposed) to be ineligible for definition in both these theories. It is also doubtful whether it can be defined in any theory comprising them. As often happens, we are left with some intuitive meaning and some rules for using the term 'existence'. Second, as a result of this, although the investigated assumptions concerning the relationship between existence and past, present, and future are formulated as equivalences, they do not possess the character of definitions of existence. They only express an extension relation between the notion of existence and notions of past, present, and future. The precise character of this relation depends on certain assumptions.

Finally, let us note that the notion of existence used here so far is also that used in existential statements (i.e., statements with an existential quantifier). This notion, as is well known, is built into the apparatus of logic, and, consequently, into set theory and the whole of mathematics. The notion of existence to be used here in the context of past, present, and future, differs from the logical notion, and is thus associated with differing intuitions. The troublesome problem of how these two notions are related will not be considered here. We shall also not enter into polemics about the opinion that only this 'logical' notion is meaningful and entirely covers the meaning of the term 'existence'.

After these introductory remarks, the above-mentioned assumptions can be formulated within standard $P^uN^uF^u$ theory,

Past, present, and future, i.e., the sets P^u_x, N^u_x, F^u_x, are doubly relativized in this theory: they are related to a specific event and to a specific inertial reference system so that the set of all existing events must also be related to a specific event and to a specific reference system. This set is denoted by E^u_x, and the formula $y \in E^u_x$ reads: the event y exists relatively to the event x in the system u. The above relativization is necessary in order to allow one to formulate the assumptions concerning the relationship between existence and past, present, and future. However, we will not try to explain in detail the meaning of the sentence 'y exists relatively to x (in the system u)'.

Different authors on the philosophy of time adopt various assumptions concerning the relationship between existence and past, present, and future. The majority of them, however, accept the commonsense view that only present events, i.e., these occurring 'now', exist. According to the terminology adopted above this assumption is formulated as follows:

(Z*) $E^u_x = N^u_x$.

In words: the event y exists with respect to the event x in the system u iff y occurs in the present of x in the system u, i.e., y is simultaneous with x in u.

Many interesting consequences follow in $P^uN^uF^u$ theory from (Z*). Here is the first one:

(Z*1) $\overline{E^u_x} = P^u_x \cup F^u_x$. {D3, D4, D5, Z*}

In words: the event y does not exist with respect to the event x in the system u iff y occurs in the past or in the future of x in the system u. The second consequence is as follows:

(Z*2) $\bigwedge_u \bigwedge_x \bigwedge_y [y \in E_x^u \equiv x \in E_y^u]$. {T21, Z*}

In words: the event y exists with respect to the event x in the system u iff x exists with respect to y in the system u. This theorem expresses the symmetry of existence under the assumption (Z*).

Two more important theorems also follow. The first is:

(Z*3) $\bigwedge_u \bigwedge_x \bigvee_y [x \in E_y^u]$. {T24, Z*}

In words: every event x exists with respect to a certain event y in the system u. The second is as follows:

(Z*4) $\bigwedge_u \bigwedge_x \bigvee_z [x \notin E_z^u]$. {T27, Z*}

In words: every event x does not exist with respect to a certain event z in the system u. Both these consequences taken together express the 'event-like' relational character of the existence of events, i.e., the fact that a given event exists with respect to some other event in a given system and does not exist with respect to another event in the same system (the system is fixed). This is obviously a universal property since it concerns every event.

The system-related relational character of the existence of events within $P^uN^uF^u$ theory under the assumption (Z*) will now be considered. Let u_1 and u_2 be two different inertial reference system moving with respect to each other. Since $u_1 \neq u_2$, we obtain for any given x: $N_x^{u_1} \neq N_x^{u_2}$, $P_x^{u_1} \neq P_x^{u_2}$, $F_x^{u_1} \neq F_x^{u_2}$. One paradoxical consequence of this state of affairs was described in Chapter 2 by the theorem (i) which is now formulated in a slightly modified and simpler form:

(i′) $\bigwedge_x \bigvee_y \bigvee_{u_1, u_2} [y \in N_x^{u_1} \wedge y \notin N_x^{u_2}]$.

In words: for every event x there exists such an event y and two different systems u_1 and u_2 such that y occurs in the present of x in u_1 and does not occur in the present of x in u_2.

In connection with the assumption (Z*), we obtain: $E_x^{u_1} = N_x^{u_1}$ and $E_x^{u_2} = N_x^{u_2}$. This fact and the theorem (i′) bring about the so-called *paradox of existence* in $P^uN^uF^u$ theory which was formulated by H. Putnam (Putnam, 1967). It reads as follows:

(ii′) $\bigwedge_x \bigvee_y \bigvee_{u_1, u_2} [y \in E_x^{u_1} \wedge y \notin E_x^{u_2}]$.

In words: for every event x there exists such an event y and two different systems u_1 and u_2 such that y exists with respect to x in the system u_1 and does not exist with respect to x in the system u_2. In short, a given event exists with respect to some event in one system and does not exist with respect to the same event in another. In order to make the theorem (ii′) a thesis of the P^{u}N^{u}F^u theory (under the assumption (Z*)) it is necessary to prove the theorem (i′) which requires completion of the theory.

The above paradox of existence expresses the purely system-related relational character of the existence of events since the reference to different events (fixed x!) is not taken into account. This is analogous to the theorems (Z*3) and (Z*4), which express the purely event-like relational character of the existence of events, since the reference of a given event to different reference systems is not taken into account. There are, of course, mixed cases, where references both to different events and to different systems take place. Consequently, the existence of an event depends at the same time on both event and system.

The alternative assumption, concerning the relationship between existence and past, present, and future within the P^{u}N^{u}F^u theory, asserts the following:

(Z′*) $E'^u_x = P^u_x \cup N^u_x \cup F^u_x$.

In words: the event y exists with respect to the event x in the system u if y occurs in the past or present or future of x in u. In short, not only present but also past and future events exist. This assumption was introduced by H. Putnam (Putnam, 1967) to eliminate the paradox of existence (ii′) which he had previously formulated. It will therefore be called *Putnam's assumption.*

Within the P^{u}N^{u}F^u theory, the assumption (Z′*) implies an important consequence which expresses the essential difference between (Z′*) and the former assumption (Z*):

(Z′*1) $\bigwedge_u \bigwedge_x \bigwedge_y [y \in E'^u_x]$.

In words: every two events exist with respect to each other in any system. Thus, all events exist. This can also be formulated as $E'^u_x = \mathbf{S}$ (it follows also from (T13) and (Z*)). This assumption therefore eliminates the double relativization of the existence of events—to events and to reference systems—for the following reasons. First, if $E'^u_x = \mathbf{S}$ for any x, then, for the same u and for $z \neq x$, we obtain $E'^u_x = E'^u_z$, i.e., for every arbitrary y, $y \in E'^u_x \equiv y \in E'^u_z$. Therefore, the existence of y with respect to x does not depend on x, thus the existence of y does not depend on events. Second, if $E'^u_x = \mathbf{S}$ for any u, then for the same x and for $u_1 \neq u_2$, we obtain $E'^{u_1}_x$

$= E_x'^{u_2}$, i.e., for every arbitrary y, $y \in E_x'^{u_1} \equiv y \in E_x'^{u_2}$. Consequently, the existence of y with respect to x does not depend on u, and thus the existence, of y does not depend on an inertial reference system. The paradox of existence which arises in the $P^uN^uF^u$ theory under the assumption (Z*) is immediately eliminated. More precisely, the paradox vanishes if (Z'*) is accepted instead of (Z*).

Two conclusions follow from these two points. The first implies that the set $E_x'^u$ is no longer related to any event, so the event-index can be omitted and we can write simply E'^u. The second implies that the set $E_x'^u$ is no longer related to an inertial reference system, so the system-index can be omitted, and we can write simply E_x'. To summarize, we can write E', i.e., omitting both indices. Hence, the expressions $x \in E'$, $y \in E'$, etc. are meaningful because, according to (Z'*1), they are equivalent to the expressions $x \in \mathbf{S}$, $y \in \mathbf{S}$, etc. Let us note that this general conclusion does not eliminate the double relativization (to events and to systems) of the sets P_x^u, N_x^u, F_x^u.

Let us now consider the comparison and evaluation of the above assumptions (Z*) and (Z'*) within the $P^uN^uF^u$ theory. It will be readily seen that the sets of existing events corresponding to these two assumptions differ. Moreover, it is clear that the set corresponding to assumption (Z*) is the smallest, and is equal to N_x^u, while the set corresponding to assumption (Z'*) is the largest, and is equal to $P_x^u \cup N_x^u \cup F_x^u$, so that it is equal to the set of all events S. This immediately implies that the former is a proper subset of the latter; in other words, if x exists with respect to y in the system u, according to the assumption (Z*), then it exists with respect to y in the system x, according to the assumption (Z'*).

Some unwelcome consequences of the assumption (Z*) may readily be noted. Firstly, it implies the event-like relativity of existence and, above all, the system-relativity of existence; the latter is expressed in the system paradox. The relativity of existence, mainly the system relativity, conflicts with the common connotation of the notion of existence of events. Second, the assumption (Z*) implies the following paradox: if $W_u(y, x)$ holds, then according to this assumption y does not exist with respect to x, and vice versa. Such an implication cannot possibly be accepted. As we have seen, the assumption (Z'*) has neither the first nor the second of these shortcomings. Still, it certainly lacks an intuitive sense. In this regard, it seems that in reality only present events exist, while past and future ones do not, and this is what (Z*) claims. Regardless of that, and following Putnam (1967), I shall choose the assumption (Z'*). In this context, we shall return to this problem when considering the evaluated assumptions in their different version, within the PNF theory.

Past, present, and future, i.e., the sets P_x, N_x, F_x, are only once relativized in this non-standard theory; they are related only to a certain, determined event. This stems from the fact that they are defined by means of the absolute time relations W, R, $\breve{W}$. Thus, the set of all existing events must also be related only to a certain event. Accordingly, this set is denoted by E_x and the formula $y \in E_x$ reads: the event y exists with respect to the event x. Such a relativization of this set is necessary to formulate various assumptions concerning the relationship between existence and absolute past, present, and future. It does not contradict the fact that one of these assumptions eliminates this event-like relativization; we have already seen this in the case of assumption (Z′*) in the $P^uN^uF^u$ theory.

At this point, besides the assumptions which could be called 'temporal', which are isomorphic to (Z*) and (Z′*) discussed in the $P^uN^uF^u$ theory, two other assumptions called 'time-causal' will also be considered. The 'temporal' assumptions will be analysed first. The commonsense belief that only present events exist is described within the PNF theory by

(Z) $E_x = N_x$.

In words: the event y exists with respect to the event x iff y occurs in the present of x, i.e., when y is quasi-simultaneous with x. Let us note that the relationship $E_x^u \subset E_x$ holds; this is a consequence of the assumptions (Z*), (Z), and the known formula $N_x^u \subset N_x$. Therefore, if x exists with respect to y in the system u, then x exists with respect to y; the reciprocal relationship obviously does not hold.

The assumption (Z) implies some consequences within PNF theory. The first is the following:

(Z1) $\overline{E}_x = P_x \cup F_x$. {D3, D4, D5, Z}

In words: the event y does not exist with respect to the event x iff y occurs in the past or future of x. This theorem is analogous to (Z*1). The second consequence is

(Z2) $\bigwedge_x \bigwedge_y [y \in E_x \equiv x \in E_y]$. {T23, Z}

In words: every event y exists with respect to the event x iff x exists with respect to y. This theorem expresses the property of symmetry of the existence of events.

Let us formulate another two fundamental theorems. The first is

(Z3) $\bigwedge_x \bigvee_y [x \in E_y]$. {T26, Z}

In words: every event x exists with respect to a certain event y. The second is

(Z4) $\bigwedge_x \bigvee_z [x \notin E_z]$. {T29, Z}

In words: every event x does not exist with respect to a certain event z. The theorems (Z3) and (Z4) together express the 'event-like' relativity of the existence of events. This means that the same event exists with respect to a certain event and does not exist with respect to some other event. This is a universal property because it concerns every event.

The most interesting consequence of the assumption (Z), within the PNF theory, is the paradox of existence. It is directly implied by the paradox of present, i.e., the theorem (T17), and by the assumption (Z). The paradox of existence is expressed by the following theorem:

(Z5) $\bigwedge_x \bigwedge_y \{W(x, y) \rightarrow \bigvee_z [x \in E_z \wedge y \in E_z]\}$. {T17, Z}

In words: for any two events x and y, if x is absolutely earlier than y, there exists such an event z that both x and y exist with respect to the event z. (T17) and (Z) also imply a theorem which differs from (Z5) only in the following way: $\breve{W}(x, y)$ is substituted for $W(x, y)$ in its antecedent. The following theorem is also important:

(Z6) $\bigwedge_x \bigwedge_y \{W(x, y) \rightarrow \bigvee_v \sim [x \in E_v \wedge y \in E_v]\}$. {T18, Z}

In words: for any two events x and y, if x is absolutely earlier than y, there exists such an event v that x and y together do not exist with respect to the event v. This happens when x occurs in the past of v (then it does not exist with respect to v), and y occurs in the future of v (then it does not exist with respect to v). (T18) and (Z) also imply the theorem, which differs from (Z6) only in the following way: $\breve{W}(x, y)$ is substituted for $W(x, y)$ in its antecedent.

The following description is a vivid illustration of the paradox of existence. Let us assume conditions similar to those which illustrated the paradox of present (see Chapter 3): $W(x, y)$ and $x, y \in a$, where a is a certain thing, and x and y are its beginning and end in time. Then, according to (Z5), there exists such an event z that the whole thing a, i.e., the whole history of a, exists with respect to z! This may be written as $a \subset E_z$. Let us return to the example of the Earth including the human race. According to the conclusion, there exists such an event z that the whole Earth, including humanity with its whole history, exists with respect to z! This event must, as we know, be very distant spatially from the Earth, something that can be calculated in STR. Let us add that according to (Z6) there also exists an event v such that it is not true that the whole Earth, including humanity with its history, exists with respect to v.

It should be noted that if the assumption (Z) is accepted, the paradox of existence follows from the paradox of present (T17), while the latter follows from the axiom (A10): $N_x \cap N_y \neq \emptyset$. The axiom (A10) is based on the non-transitivity of the quasi-simultaneity relation. This fact substantially differentiates the PNF theory from the $P^uN^uF^u$ theory. The quasi-present N_x can be therefore extended in time, and the paradox of existence (under the assumption (Z)) becomes comprehensible even though it remains inconsistent with common intuition.

The second assumption (isomorphic to (Z′*)) concerning the relationship between the sets P_x, N_x, F_x and the set E'_x states:

(Z′) $E'_x = P_x \cup N_x \cup F_x$.

In words: the event y exists with respect to the event x iff y occurs in the past, present, or future of x. In short: not only present but also past and future events exist. This assumption will be called—like (Z′*)—*Putnam's assumption*. These are, simply, two different versions corresponding to two different theories: the former to the $P^uN^uF^u$ theory, the latter to the PNF theory. It should be noted that in this case $E'^u_x = E'_x$, since the sets $P^u_x \cup N^u_x \cup F^u_x$ and $P_x \cup N_x \cup F_x$ are equal. Hence, if x exists with respect to y, in the system u, then x exists with respect to y, and vice versa.

In the framework of PNF theory, the assumption (Z′) implies an important consequence which determines the essential distinction between (Z′) and (Z):

(Z′1) $\bigwedge_x \bigwedge_y [y \in E'_x]$. {T13, Z′}

In words: every two events exist with respect to each other, i.e., all events exist. This can be also expressed by the equality $E'_x = \mathbf{S}$. The assumption (Z′) eliminates the event-like relativization of the existence of events. This is the case because if $E'_x = \mathbf{S}$ (for any x), then, given that $z \neq x$, we obtain $E'_x = E'_z$, i.e., $y \in E'_x \equiv y \in E'_z$. Consequently, the existence of y with respect to x does not at all depend on x. Hence, the set E'_x is no longer related to any event, and the event-index may be omitted, so we can write simply E'. Therefore, the expressions $x \in E', y \in E'$, etc. are meaningful, as they are equivalent to the expressions $x \in \mathbf{S}$, $y \in \mathbf{S}$, etc. It should be added that the consequence considered above clearly does not eliminate (on the basis of (Z)) the reference of the sets P_x, N_x, F_x to events.

It is very important that the assumption (Z′) also eliminates the paradox of existence which appears in the PNF theory, under the assumption (Z). This elimination does not consist in the fact that, in the face of (Z′), the theorem (Z5) is no longer true. It can be easily demonstrated (provided it is noted that, according to (Z′), the relationship $N_x \subset E'_x$ holds). The point is that in this case the theorem (Z5) ceases to be paradoxical. This means that if $W(x, y)$ (antecedent of (Z5)) holds, then, according to (Z′), both events x and y exist not only with respect to a certain event z, in the present of which they occur, but they also exist with respect to any event in whose past or future, for example, they occur. This is because, according to (Z′1), every two events exist with respect to each other or, simply, every event exists!

It should be added that the system-related paradox of existence, which appears in the $P^u N^u F^u$ theory under the assumption of (Z*) (analogous to (Z)), is completely different from the paradox under consideration and comes from a different source. Still, we know it can be eliminated if, analogously to (Z′), (Z′*) is assumed. Thus the paradoxes are different, but the method of elimination remains the same.

It is now time to introduce the previously mentioned assumptions concerning the relationship between the set E_x and the sets P_x, N_x, F_x which differ from both (Z) and (Z′). They are called 'time-causal' because they contain, apart from temporal concepts, the symbol H which denotes a causal relation, i.e., the relation of physical interaction. It should be pointed out immediately that these assumptions will be considered only within PNF theory, because the conception of the relation H has been developed in this theory alone.

According to this idea, the relation H, which is defined on the set S, is asymmetric and transitive there, and is thus irreflexive ((A11), (A12), (T31) from Chapter 3); the converse of H, i.e., the relation $\breve{H}$ which is also obviously absolute, possesses in S the same property. Let us recall the so-called postulate of causality also assumed in the Chapter 3: $H \subset W$, and its consequence $\breve{H} \subset \breve{W}$ ((A15), (T17) from Chapter 3).

The first assumption is due to J. Łukasiewicz (Łukasiewicz, 1961), while the second, simpler assumption will be now formulate by me. The latter will be analysed first to enable a better understanding of the former. After the universal quantifiers for event variables are omitted, the assumption considered takes the form:

(ZH1) $y \in E^1_x \equiv \{[y \in N_x] \vee [y \in P_x \wedge H(y, x)] \vee [y \in F_x \wedge \breve{H}(y, x)]\}.$

In words: the event y exists with respect to the event x iff y occurs in the present of x, or y occurs in the past of x and is the cause of x, or y occurs in the future of x and is the result of x. The components of this threefold disjunction obviously exclude each other. This can be easily demonstrated on the light cone $C(x)$ of the event x.

The assumption (ZH1) has a shorter equivalent form, which is as follows:

$$y \in E_x^1 \equiv [y \in N_x \vee H(y, x) \vee \breve{H}(y, x)].$$

In words: the event y exists with respect to the event x iff y occurs in the present of x, or y is a cause of x, or y is a result of x; this version seems to be clearer than (ZH1). Proof: according to the definitions of the sets P_x and F_x (D4, D5), the postulate of causality and its consequence have the forms: $H(y, x) \rightarrow y \in P_x$ and $\breve{H}(y, x) \rightarrow y \in F_x$. The above theorem is obtained by virtue of these theses and the following tautology: $[(p \rightarrow q) \rightarrow (p \wedge q \equiv p)]$.

The second assumption, that of Łukasiewicz (1961), is formulated here not in its original version (non-formalized, non-relativistic and employing the notion of moment), but within the framework of the PNF theory. After the universal quantifiers are omitted, as before, the assumption takes the form:

$$\text{(ZH2)} \quad y \in E_x^2 \equiv \Big\{y \in N_x \vee y \in P_x \wedge \bigvee_z [z \in N_x \wedge H(y, z)] \vee$$
$$\vee\, y \in F_x \wedge \bigvee_v [v \in N_x \wedge \breve{H}(y, v)]\Big\}.$$

In words: the event y exists with respect to the event x iff (1) y occurs in the present of x, or (2) y occurs in the past of x and is a cause of a certain event z which occurs in the present of x, or (3) y occurs in the future of x and is a result of a certain event v which occurs in the present of x. The components of this threefold disjunction are mutually exclusive. They may be easily demonstrated on the light cone $C(x)$ of the event x. It is worth emphasizing that Łukasiewicz was not the only Polish philosopher of science to put forward this assumption, obviously in its original version. It was also accepted by T. Kotarbiński among others (Kotarbiński, 1913).

I am of the opinion that the existence of events, according to both assumptions (ZH1) and (ZH2), is relational in its character (analogously to (Z)). This means that in the face of these assumptions (each one taken separately) the theorems, analogous to the consequences (Z3) and (Z4) of the assumption (Z), are true. We do not, however, present a proof of these theorems in PNF theory based on these assumptions. However, I suppose such proofs are possible, because, according to the contents of the assumptions, the sets of the type E_x corresponding to these assumptions are proper subsets of the union $P_x \cup N_x \cup F_x$. The discussed (event-like) relational character of the existence of events is, as we know, eliminated only under the assumption (Z′) (when $E'_x = P_x \cup N_x \cup F_x$). It should nonetheless be emphasized that the existence of events under Łukasiewicz's (i.e., (ZH2)) assumption has, as we shall soon see, a relational character, provided that a certain principle, called the *principle of strict determinism*, is refuted.

The relationship between the sets E_x, E_x^1, E_x^2 and E'_x, i.e., between the notions of existence, related to the assumptions presented, will now be investigated. As we shall see, it suffices to describe the relationships between the elements of the following pairs: E_x and E_x^1, E_x^1 and E_x^2, E_x^2 and E'_x, in order to determine the relationships between the elements of all remaining pairs.

First, let us consider the pair E_x and E_x^1; in this case, the following theorem is true:

(1) $E_x \subset E_x^1$.

In words: if the event y exists with respect to the event x, according to the assumption (Z), then the event y exists with respect to the event x, according to the assumption (ZH1). Proof: according to (Z) and (ZH1), the following relationships hold: $E_x \subset N_x$ and $N_x \subset E_x^1$, and they imply the theorem (1).

In case of the pair E_x^1 and E_x^2, the following theorem is true:

(2) $E_x^1 \subset E_x^2$.

In words: if the event y exists with respect to the event x, according to the assumption (ZH1), then the event y exists with respect to the event x, according to (ZH2). Proof: the following two implications can be demonstrated:

$$H(y, x) \rightarrow \{y \in P_x \wedge \bigvee_z [z \in N_x \wedge H(y, z)]\},$$

$$\breve{H}(y, x) \rightarrow \{y \in F_x \wedge \bigvee_v [v \in N_x \wedge \breve{H}(y, v)]\}.$$

The implication $y \in N_x \rightarrow y \in N_x$ is self-evident. The summing up of these three implications gives such the implication that its antecedent is the right-hand side of the assumption (ZH1) and its consequent is the right-hand side of the assumption (ZH2)—therefore the theorem (2) is obtained.

Finally, in the case of the pair E_x^2 and E'_x, the following statement is true:

(3) $E_x^2 \subset E'_x$.

In words: if the event y exists with respect to the event x, according to the assumption (ZH2), then the event y exists with respect to the event x, according to the assumption (Z′). Proof: the following implications are self-evident:

$$y \in N_x \rightarrow y \in N_x,$$

$$\{y \in P_x \wedge \bigvee_z [z \in N_x \wedge H(y, z)]\} \rightarrow y \in P_x,$$

$$\{y \in F_x \wedge \bigvee_v [v \in N_x \wedge \breve{H}(y, v)]\} \rightarrow y \in F_x.$$

The summing up of these gives such the implication that its antecedent is the right-hand side of the assumption (ZH2) and its consequent is the right-hand side of (Z′)—thus the theorem (3) is obtained.

Consequently, we have the situation expressed by the global statement:

(∗) $E_x \subset E_x^1 \subset E_x^2 \subset E_x'$.

Due to the transitivity of the inclusion $\subset$, the theorems (1), (2), (3) imply the following theorems concerning the remaining pairs:

(4) $E_x \subset E_x^2$,

(5) $E_x \subset E_x'$,

(6) $E_x^1 \subset E_x'$.

The established theorems indicate that if we pass from the assumption (Z), through the assumptions (ZH1) and (ZH2), to the assumption (Z′), we encounter an ever-broadening set of existing events. The set E_x is the smallest one according to the assumption (Z)—it is equal to the set N_x. Hence, no events from the set P_x and F_x belong to this set. According to the assumption (ZH1), the set E_x^1 is larger than the set E_x. It contains—besides the set N_x—also some subsets of the sets P_x and F_x, elements of which together with the event x satisfy the relation H or $\breve{H}$. According to the assumption (ZH2), the set E_x^2 is larger than the set E_x^1. Beside the set N_x, it also contains some subsets of the sets P_x and F_x, some elements of which bear the relation H or $\breve{H}$ to the event x, while some other elements bear these relations to elements from the set N_x. Finally, according to the assumption (Z′), the set E_x' is the largest; it is equal to the sum of the sets P_x, N_x, F_x, i.e., to the set of all events S.

From these considerations it follows that (ZH1) and (ZH2) are assumptions intermediate between the minimal assumption (Z), and the maximal assumption (Z′). These 'time-causal' assumptions explicitly impose restrictions of a causal kind on the sets P_x and F_x. The restrictions are stronger in the case of (ZH1) and weaker in the case of (ZH2). As a result, they distinguish within these sets smaller or larger subsets which set up the sets E_x^1 and E_x^2. The following two facts deserve emphasis at this point. First, according to all these assumptions, different sets of existing events always contain the set N_x. Secondly, according to the intermediate assumptions (ZH1) and (ZH2), these sets contain only such subsets of P_x and F_x that their elements are causally connected with elements of the set N_x. Hence, the set N_x evidently holds a distinctive position with respect to existence.

The following question arises in this connection: on what grounds is such a position of the present based? It is difficult to find a satisfactory answer to this question even among claims of adherents to assumption (Z) (cf. Gale, 1968) who, after all, reduce existing events to present events. The answer should probably be sought in the very meaning of the notion of existence which has been implicitly accepted here, or perhaps in the criteria of its application; although these are only conjectures.

Another question arises here concerning the assumptions (ZH1) and (ZH2) alone. When going beyond the present (N_x) in these assumptions, i.e., extending existence (E_x) to a certain part of the past (P_x) and future (F_x), why do we proceed via the causal connection? An answer to this question cannot even be found in the writings of authors who—like Kotarbiński (1913) and Łukasiewicz (1961)—although not the originators of the idea of a connection between causality and existence, themselves went on to develop it. Perhaps I have not been sufficiently persistent in looking for the answer. This is undoubtedly a fundamental problem, but one that I am not at present capable of solving; for this reason, I have only hinted at it.

The results expressed in the global theorem (∗), as has already been pointed out, demand some sort of qualification concerning the connection $E_x^2 \subset E_x'$. Namely, in the face of a certain causal principle, the connection between E_x^2 and E_x' is stronger than the one declared. This is the principle of strict determinism, or strict causality, which is a conjunction of two logically independent theorems. The original version of this principle by J. Łukasiewicz (Łukasiewicz, 1961) (non-formalized, disregarding STR, and using the notion of moment) can be formulated in the conceptual framework of PNF theory as follows:

$$\text{(PSD1)} \quad \bigwedge_y \bigwedge_x \{W(y, x) \rightarrow \bigvee_z [R(z, x) \wedge H(y, z)]\}.$$

In words: every event y earlier than any event x is a cause of some event

z quasi-simultaneous with x. In other words: every event has a result in the present of any event, in the past of which it occurs.

$$\text{(PSD2)} \quad \bigwedge_{y}\bigwedge_{x}\{\breve{W}(y, x) \rightarrow \bigvee_{v} [R(v, x) \wedge \breve{H}(y, v)]\}.$$

In words: every event y later than any event x is a result of some event v quasi-simultaneous with x. In other words: every event has a cause in the present of any event, in the future of which it occurs. Both theorems obviously are logically independent.

It is not difficult to prove that, if the principle of strict determinism is assumed, then the following theorem is true:

(3′) $E_x^2 = E'_x$.

In words: the event y exists with respect to the event x, according to the assumption (ZH2), iff the event y exists with respect to the event x, according to the assumption (Z′). Proof:

(i) (PSD1) is equivalent to the theorem:

$$y \in P_x \rightarrow \bigvee_{z} [z \in N_x \wedge H(y, z)],$$

which, by virtue of the tautology $[(p \rightarrow q) \rightarrow (p \wedge q \equiv p)]$, implies the following equivalence:

$$\{y \in P_x \wedge \bigvee_{z} [z \in N_x \wedge H(y, z)]\} \equiv y \in P_x.$$

The left-hand side of this equivalence is the second part of the right-hand side of the assumption (ZH2), and its right-hand side is the second part of the right-hand side of the assumption (Z′).

(ii) (PSD2) is equivalent to the theorem:

$$y \in F_x \rightarrow \bigvee_{v} [v \in N_x \wedge \breve{H}(y, v)],$$

which, by virtue of the same tautology, implies the equivalence:

$$\{y \in F_x \wedge \bigwedge_{v} [v \in N_x \wedge \breve{H}(y, v)]\} \equiv y \in F_x.$$

The left-hand side of this equivalence is the third part of the right-hand side of the assumption (Z′).

(iii) The first parts of both assumptions are the same, i.e., $y \in N_x$. Hence we obtain theorem (3′).

Theorem (3′) immediately implies—provided that (PSD1) and (PSD2) are accepted—that assumptions (Z′) and (ZH2) are equivalent. In this case, the existence of events, according to the assumptions (ZH2), does not possess a relational character, i.e., any two events exist with respect to each other. This equivalence has not previously been acknowledged. The reason for this is probably the fact that these assumptions have never been compared in the light of the principle of strict determinism. This principle and the assumptions (ZH2) imply the assumption (Z′), as was demonstrated earlier.

Łukasiewicz categorically rejected the principle of strict determinism as being doubtful from an empirical point of view, as well as from that of an intuitive perspective, and I would agree with him.

As a consequence of such a rejection, it is no longer possible to causally justify the assumption (Z′). As we have seen, (Z′) is equivalent to (ZH2) only if the principle of strict determinism is presupposed. Hence, if the principle is rejected, a different justification of (Z′) has to be sought.

It is high time to proceed to a critical evaluation of the assumptions formulated above concerning the relationship between existence and past, present, and future within PNF theory. Let us note that these do not represent all possible assumptions of this kind. Some others, besides (Z) and (Z′), can be formulated within the class of 'temporal' assumptions, for instance the following: $E_x = N_x \cup P_x$, i.e., only present and past events exist. This last assumption (disregarding STR) was accepted, for instance by C. D. Broad (Broad, 1923). The same is true of the class of 'time-causal' assumptions. Only four of them—those which seem to me be rational and plausible—have been analysed here.

The evaluation will first of all indicate those consequences of the assumptions which I consider to be shortcomings. The first three, i.e., (Z), (ZH1), and (ZH2), imply the event-like relational character of existence. This has been proved in the case of the assumption (Z), as well as in the case of the assumptions (ZH1) and (ZH2). (In the case of (ZH2), the principle of strict determinism had to be refuted.) This implication is regarded as an important defect of the assumptions. This relational character seriously conflicts with the commonly accepted connotation of the notion of existence of any objects whatsoever, including events. This view requires justification, which is, however, not possible here since it presupposes a very complex analysis of the meaning of the term 'existence'.

Secondly, the assumption (Z) implies the paradox of existence (i), which is very troublesome not only from the point of view of intuition. Thirdly, (Z) entails a well-known paradox: if $W(y, x)$ takes place, then, according to (Z), the event y does not exist with respect to the event x, and the event y does not exist with respect to the event x. This consequence is rather shocking.

The above-mentioned paradox is not so drastic in the case of the assumptions (ZH1) and (ZH2). If the conjunction $W(y, x) \wedge H(y, x)$ holds, then, according to (ZH1), the event y exists with respect to the event x, and vice versa. Whereas, while the conjunction $W(y, x) \wedge \bigvee_z [R(z, x) \wedge H(y, z)]$ holds, then, according to (ZH2), the event y exists with respect to event x, and—obviously—vice versa. A weaker additional condition suffices here, when compared with that of (ZH1); events for which W holds exist with respect to each other. Nevertheless, there are such events for which W holds which do not exist with respect to each other under both the assumptions.

Finally, the assumption (Z) also implies the following paradox: if $H(y, x)$ holds, then the event y does not exist with respect to the event x, i.e., $H(y, x) \rightarrow y \notin E_x$. This theorem follows from the assumption (Z) and the postulate of causality (A15). Let us note that under both the intermediate assumptions (ZH1) and (ZH2), and when the postulate (A15) is taken into account, the following theorem is true: $H(y, x) \rightarrow y \in E_x$. Therefore, in neither of these cases does this paradox occur.

The difficulties presented here indicate that of the three assumptions (Z), (ZH1) and (ZH2), the most doubtful is (Z). It possesses the greatest number of defects. Given the fact that it is the most intuitive of all four assumptions, this is very strange—not to say paradoxical. It is generally accepted as something fairly obvious that only present events exist, while past and future ones do not: the former no longer exist, the latter not yet. This belief has acquired the status of a philosophical principle.

Intuition, however, cannot be accepted as a basic criterion of choice. The criterion can only be the number of difficulties following from the assumption. According to this criterion, I reject the assumption (Z) first. The assumptions (ZH1) and (ZH2) are also rejected, though with less conviction. The assumption (Z′), even though it is inconsistent with intuition, is accepted. It is the most liberal assumption since it admits the maximal set of existing events. It possesses none of the weaknesses which characterize the remaining assumptions. First of all, it does not imply the (event-like) relational character of the existence of events: it entails, as has been already shown, the theorem (Z′1) which claims that every two events exist with respect to each other, i.e., that all events exist. At this point the problem arises whether, according to (Z′), the notion of existence, which is used here, does not have the same extension as the 'logical' notion of existence. This remains, however, outside the range of the present work.

Only one objection—except the contradiction of commonsense—is often raised against the chosen assumption. Namely, it seems to imply that all events are present ones, i.e.—in terms of PNF theory—that $\mathbf{S} \subset N_x$. This is, however, a delusion which stems from the misunderstanding of the content of the assumption and its consequences. The assumption (Z′) entails only the theorem: $N_x \subset E'_x$, not the reciprocal theorem: $E'_x \subset N_x$. The latter, together with the genuine consequence of (Z′): $E'_x = \mathbf{S}$, obviously imply $\mathbf{S} \subset N_x$. This absurd statement is not, however, a consequence of (Z′), since $E'_x \subset N_x$ is not its consequence. As we can see, proponents of this objection not only do not understand the assumption (Z′), but they are also ignorant of elementary logic.

To conclude the present chapter, it should be emphasized that the analysis of the relationship between existence and past, present, and future which has been carried out here has barely touched on this enormously complex problem. This complexity stems most of all from the presence of the notion of existence. This analysis has been carried out in a quite restrictive manner—within the framework of STR and the two systems formulated here: $P^uN^uF^u$ theory and PNF theory. Nonetheless, these very limitations allow for a relatively clear consideration of the subject.

CHAPTER 6

PAST, PRESENT, FUTURE AND BECOMING

The final problem which deserves attention here is the connotation of such notions as *coming-to-be*, *lasting* or *duration*, and *passing away*, and the notion of the *becoming* of things as well as events. We will attempt to define these notions, but only by means of the notions of past, present, and future used so far. This approach implies, first, that these notions will only have a temporal character, and secondly, that they will be of an explicitly relational kind—by analogy to the notions of past, present, and future. Of course, we shall not confine ourselves only to these definitions but shall also attempt to formulate—as far as this is possible—certain theorems concerning these notions, their mutual connections, and their association with existence.

The problem will be discussed within the framework of PNF theory only. This means that some specific features will be imposed on the defined notions, namely the features associated with PNF theory, and this special aspect will be emphasized whenever it appears to be of particular importance or interest. We shall see that the merits of adopting such a framework are substantial. The first advantage is the relative simplicity of definitions. The second and essential one is that we shall give definitions of the absolute (that is independent of the inertial reference system) notions of the coming-to-be, lasting, passing away, and becoming of things and events. It is always possible to discuss this problem within $P^uN^uF^u$ theory, relying on the results of the present chapter.

We begin with the pair of notions of coming-to-be and passing away. The binary relation A of coming-to-be (creation) is introduced into the set **T** of things interpreted in an event-like manner. The formula $A(a|b)$ reads: the thing a comes-to-be with respect to the thing b. The definition of the relation A is as follows (universal quantifiers binding the thing-variables are omitted):

(D1) $A(a|b) \equiv \bigvee_{y \in b} \bigwedge_{x \in a} [x \in F_y] \wedge \bigvee_{y \in b} \bigvee_{x \in a} [x \in N_{y'}]$.

In words: the thing *a* comes-to-be with respect to the thing *b* iff (i) there is such an event *y* from *b* that every event *x* from *a* belongs to the future of *y*, and (ii) there is such an event *y'* from *b* and such an event *x* from *a* that *x* belongs to the present of *y'*. The expression '*x* from *a*' is an abbreviation of '*x* is an element of *a*' (since *a* is a set of events) and will be frequently used hereafter. From the intuitive point of view, the first constituent of the definiens in the definition of the relation *A* is obvious—one can say that the thing *a* comes-to-be with respect to the thing *b* only if the entire *a* belongs (in time) to the future of a certain event from *b*, i.e., of a certain time region of *b*. Let us at once note that this definition does not assume that things possess their beginnings in time. All that is assumed is that things are (metrically) limited in time. It is worth adding perhaps that if we use the notion of the beginning of the thing in time, it is easy to define the non-relational coming-to-be of things. Namely: 'the thing comes-to-be' means that the thing *a* has its beginning in time. How to define the coming-to-be of the thing in a non-relational way, yet with no reference to a beginning in time, is itself an interesting question.

The status of the second constituent of the definiens in the definition of the relation *A* is somewhat doubtful. Intuition seems to demand this component; in order to say that the thing *a* comes-to-be with respect to the thing *b* one must also assume that certain events from *a* and *b* are quasi-simultaneous, i.e., that a certain time-part of the thing *a* belongs to the present of a certain time-part of the thing *b*. If the definition (D1) were to be restricted to its first part, then the field of the relation *A* would be extended too much. *A* would also hold for *a* and *b* if *a* was simply later than *b*—without any (partial) coincidence in time with *b*.

The structural properties of the relation *A* will now be considered. The definition (D1) of *A* and the analysis of the light cones of events from *a* and *b* provides the grounds on which these properties can be established. Firstly, the relation *A* is asymmetric in the set **T**: $A(a|b) \rightarrow \sim A(b|a)$, i.e., if *a* comes-to-be with respect to *b*, then *b* does not come-to-be with respect to *a*. Secondly, the relation *A* is irreflexive in the set **T**: $\sim A(a|a)$, i.e., no thing comes-to-be with respect to itself. This of course follows from the fact that the relation *A* is asymmetric. Thirdly, and finally, in the set **T** the relation *A* is neither transitive nor intransitive. It is easy to indicate, by a conical analysis, a situation when $A(a|b) \wedge A(b|c) \wedge \sim A(a|c)$, which would contradict the transitivity of *A*, and also a situation when $A(a|b)$

$\wedge A(b|c) \wedge A(a|c)$, which, in turn, would contradict the intransitivity of the relation A.

The binary relation Z of the passing away (annihilation) of things will now be introduced into the set **T**. The formula $Z(a|b)$ reads: the thing a passes away with respect to the thing b. The definition of the relation is as follows:

$$\text{(D2)} \quad Z(a|b) \equiv \bigvee_{z \in b} \bigwedge_{x \in a} [x \in P_z] \wedge \bigvee_{z' \in b} \bigvee_{x \in a} [x \in N_{z'}].$$

In words: the thing a passes away with respect to the thing b iff (i) there is such an event z from b that every event x from a belongs to the past of z, and (ii) there is such an event z' from b and such an event x from a that x belong to the present of z'.

The first constituent of the definiens in (D2) is intuitively obvious—one can say that the thing a passes away with respect to the thing b only if the entire a belongs (in time) to the past of a certain event from b, i.e., of a certain time region of b. The second constituent of the definiens in (D2) is also intuitively necessary; to say that the thing a passes away with respect to the thing b, one must also assume that certain events from a and b are quasi-simultaneous, i.e., that a certain time-part of the thing a belongs to the present of a certain time-part of the thing b. If (D2) were to be restricted to its first constituent, the relation Z would also hold for a and b, provided that a was merely earlier than b, with no (partial) coincidence in time with b.

It should be emphasized that the definition of the relation Z does not assume that things possess their ends in time. It is only assumed that things are (metrically) limited in time. It should be emphasized also that non-relational passing away can be defined by means of the notion of the end of the thing in time. How to define passing away with no reference to the notion of the end of the thing in time is a question worth discussing.

What are the structural properties of the defined relation Z? The answer is not difficult if it is based on the definition (D2) and conical analysis. The relation Z is therefore asymmetric in the set **T**, i.e., $Z(a|b) \rightarrow \sim Z(b|a)$. This means that if a passes away with respect to b, then b does not pass away with respect to a. Moreover, the relation Z is irreflexive in the set **T**, i.e., $\sim Z(a|a)$, i.e., no thing passes away with respect to itself; this follows from the asymmetry of Z. Finally, the relation Z is neither transitive nor intransitive. It is possible to find a situation demonstrating that $Z(a|b) \wedge Z(b|c) \wedge \sim Z(a|c)$, which contradicts the transitivity of Z as well as a situation where $Z(a|b) \wedge Z(b|c) \wedge Z(a|c)$, which, in turn, contradicts the intransitivity of the relation Z.

When the defined relations A and Z are compared, they turn out to resemble each other closely. Formally, (D2) results from (D1) after the symbol of past is substituted for the symbol of future. The second constituent of both definienses in the definitions is explicitly the same—this point will also be discussed. This constituent determines a certain new and important time notion from the class of those discussed here. Finally, both these relations possess exactly the same formal properties: they are asymmetric and therefore irreflexive.

The lasting (duration) of things is the next notion analysed here. The binary relation D of lasting of things is introduced into the set **T**. The formula $D(a|b)$ reads: the thing a lasts with respect to the thing b. The relation is defined as follows:

$$\text{(D3)} \qquad D(a|b) \equiv \bigvee_{y \in b} \bigvee_{x \in a} [y \in N_x].$$

In words: the thing a lasts with respect to the thing b if a certain event from b occurs in the present of a certain event from a, or, more simply, a certain time-part of the thing a coincides with a certain time-part of the thing b (x and y are quasi-simultaneous). This is compatible with the intuition of relative lasting—there is no relative lasting without this coincidence.

Let us note that, almost as a rule, the notion of the lasting of things is used in a non-relational sense. In my opinion, it is equivalent to the notion of the extension of things in time. The latter, in turn, can be defined as in Chapter 3, provided that things are regarded as sets of events. This means that the definition is formulated on the ground of the assertion that at least two events from the thing a stand in a relation of absolute time separation (i.e., the relation $\bar{R}$ which is equal to the sum $W \cup \breve{W}$). This non-relational notion of lasting has little in common with the relational notion of lasting defined here.

The structural properties of the relation D are as follows: the definition (D3) implies that the relation is reflexive in the set **T**, i.e., $D(a|a)$: a thing lasts with respect to itself. This definition also implies that the relation is symmetric in the set **T**, i.e., $D(a|b) \rightarrow D(b|a)$, i.e., if the thing a lasts with respect to the thing b, then the thing b lasts with respect to the thing a. Both these properties are obvious, provided that the relation of quasi-simultaneity R (which defines the set N_y and therefore the relation D) is reflexive and symmetric. Finally, the relation D is not transitive in the set **T** since it is possible to have situations where $D(a|b) \wedge D(b|c) \wedge \sim D(a|c)$. Neither it is intransitive in **T** since there occur situations when $D(a|b)$

$\wedge D(b|c) \wedge D(a|c)$. This is the consequence of the fact that the relation of quasi-simultaneity is neither transitive nor intransitive.

The relation D is included in both the relations A and Z. This becomes evident when the definition of the relation D is compared with the definitions of the relations A and Z. The following theorems can therefore be accepted:

(T1) $A \subset D,$ {D1, D3}

(T2) $Z \subset D.$ {D2, D3}

In words: if a comes-to-be with respect to b, then a lasts with respect to b (T1); if a passes away with respect to b, then a lasts with respect to b (T2); the reciprocal theorems are, of course, false.

The fourth and last temporal notion considered here is the becoming of things. The binary relation B of becoming of things is introduced into the set **T**. The formula $B(a|b)$ says that the thing a becomes with respect to the thing b. This relation can be defined, according to our understanding of its content, by means of the previously introduced relations A and Z. It is therefore determined by the sets P_x, N_x, F_x analogously to A and Z. The definition is

(D4) $B(a|b) \equiv A(a|b) \wedge Z(a|b).$

In words: the thing a becomes with respect to the thing b iff a comes-to-be and passes away with respect to b. This definition may be written more simply as $B = A \cap Z$. Because of the relationships $A \subset D$ and $Z \subset D$, one can claim that the thing a becomes with respect to b when it comes-to-be, lasts, and passes away with respect to b (lasting is here expressed explicitly).

To avoid any possible misunderstandings, some further comments on this definition are needed. It should be stressed at once that, according to the definition (D4), becoming has nothing in common with the notion of a thing 'becoming such and such', i.e., with the aquisition of properties not possessed earlier. The latter notion is associated with the transitory properties of the thing, and does not have a relational character. What is meant here is, approximately, the becoming of things that is associated only with time (and possibly with existence).

Furthermore, even if the relational approach to becoming is accepted, there may still be some doubt as to whether definition (D4) is adequate to the meaning of the term. The becoming of things is currently understood as coming-to-be, while definition (D4), it may be argued, describes rather a lapse of a thing, since the thing comes-to-be, lasts and passes away. Against this it may be argued that 'lapse' in fact means that the thing lasts and passes away. In these circumstances, we must cut the argument short and simply accept (D4) as a terminological convention. This actually was done when we formulated the definition (D4), which imposed some conventional meaning on the term 'becoming' of things. A reader who so wishes may use terms like 'termination', 'extinction', 'expiration', or even 'passing' instead. What is meant is quite clear. Lastly, two notions of becoming which do not concern things will be introduced in the subsequent part of the present chapter. They will refer to events; as we shall see, they are related to the notions considered now and are philosophically important. The problems discussed will be recalled there in a different aspect.

Let us now pose the question: what structural properties does the relation B possess? The fact that its constituent relations A and Z are asymmetric in the set $\mathbf{T}$ and the definition (D4), imply that the relation B is also asymmetric in the set $\mathbf{T}$. This means that if a becomes with respect to b, then b does not become with respect to a. Formally: $B(a|b) \rightarrow \sim B(b|a)$. The relation B is obviously irreflexive in the set $\mathbf{T}$, i.e., $\sim B(a|a)$—no thing becomes with respect to itself. This follows indirectly from the irreflexivity of the relations A and Z, and directly from the stated asymmetry of the relation B. Finally, the relation B seems to be transitive since no situation can occur in which $B(a|b) \wedge B(b|c) \wedge \sim B(a|c)$.

Let us now proceed to the problem of the connections between the relation B and the relations A, Z, and D. The definition (D4) immediately implies the following theorems:

(T3) $B \subset A$, {D4, D1}

(T4) $B \subset Z$. {D4, D2}

In words: if the thing a becomes with respect to the thing b, then a comes-to-be with respect to b, and if a becomes with respect to b, then a passes away with respect to b. The theorems (T1) and (T3), or (T2) and (T4), imply the following (obvious) theorem:

(T5) $B \subset D$. {T1, T3}

In words: if the thing a becomes with respect to the thing b, then a lasts with respect to b. To sum up: the relation B is included in every one of the three relations defined and discussed above, A, Z, and D.

The relations of coming-to-be A and passing away Z are not connected by inclusion in any way. Still, they are both included in the relation of lasting D. It should be stressed that the following four cases of product of these two relations or their complements are possible; namely, $A \cap Z$, $A \cap \bar{Z}$, $\bar{A} \cap Z$, $\bar{A} \cap \bar{Z}$. These products are, of course, mutually exclusive.

On the other hand, the constituents of each pair are consistent. Let us note that any two things stand in only one of these product relations. The four cases given therefore cover all possible cases and will now be considered in turn. As we know, the product $A \cap Z$ is equal to the relation B, i.e., to becoming of things. The product $\bar{A} \cap \bar{Z}$ means that the thing a neither comes-to-be nor passes away with respect to b. The remaining products describe the intermediate situations: the thing a comes-to-be, but does not pass away with respect to b; or, the thing a passes away, but does not come-to-be with respect to b. The last three cases exclude the relation of becoming B.

Let Q denote any of the defined relations A, Z, or B. Let us consider questions which are substitutions of the schema: 'Does each thing stand in the relation Q to a particular thing?' The axioms of PNF theory, definitions of the relations A, Z, and B, and certain assumptions which have been adopted explicitly about the extension of things in time, are not enough to allow for the proof of the affirmative answers to these questions; the answers can therefore only be assumed axiomatically. This would seem to be quite sensible on the grounds of commonsense as well as scientific experience concerning things.

It appears that it is enough to adopt just one such axiom for the relation of becoming B, namely:

(A1) $\bigwedge_a \bigvee_b [B(a|b)]$.

In words: each thing a becomes with respect to a particular thing b. Since, according to (D4), we have $B = A \cap Z$, (A1) entails the following theses:

(T6) $\bigwedge_a \bigvee_b [A(a|b)]$, {D4, A1}

(T7) $\bigwedge_a \bigvee_b [Z(a|b)]$. {D4, A1}

In words: each thing a becomes with respect to a particular thing b (T6), and each thing a passes away with respect to a particular thing b (T7). Let us note that it is not difficult to demonstrate that the axiom (A1) does not result from the conjunction of the theses (T6), (T7) and the definition (D4), though this might seem to be the case. Therefore, if (A1) were to be questioned while (T6) and (T7) were not, these theorems would have to be adopted as axioms. Thus, (A1) is a very strong theorem.

Let us now consider questions that are versions of the schema: 'Does each thing not stand in the relation Q to a certain different thing?' In this case, as in the previous one, PNF theory does not provide a basis for proving affirmative answers. The answers can only be assumed axiomatically, guided by data drawn from experience.

Two axioms must therefore be adopted: the first for the relation of coming-to-be A, and the second for the relation of passing away Z. Thus we obtain:

$$\text{(A2)} \quad \bigwedge_a \bigvee_c \{\sim [A(a|c)] \wedge c \neq a\},$$

$$\text{(A3)} \quad \bigwedge_a \bigvee_c \{\sim [Z(a|c)] \wedge c \neq a\}.$$

In words: every thing a does not come-to-be with respect to some other thing c (A2), and every thing a does not pass away with respect to some other thing c (A3). From the definition (D4) it follows that $B = A \cap Z$; this and the disjunction of the axioms (A2) and (A3) imply

$$\text{(T8)} \quad \bigwedge_a \bigvee_c \{\sim [B(a|c)] \wedge c \neq a\}. \qquad \{\text{D4, A2, A3}\}$$

In words: every thing a does not become with respect to some other thing c. Let us note at once that the axioms (A2) and (A3) do not result from this thesis and the definition (D4), though this might seem to be the case. Here we have a situation opposite to the one previously discussed. Therefore, if (A2) and (A3) were to be questioned while (T8) was not, the theorem (T8) would have to be adopted as an axiom.

Some important implications concerning the relations A, Z, and B follow from these theses and axioms. (T6) and (A2) taken together express the relational character of the coming-to-be of things, i.e., the fact that a thing may come-to-be with respect to one thing and not come-to-be with respect to another. Further, (T7) and (A3) taken together express the relational character of the passing away of things: a thing may pass away with respect to one thing and not pass away with respect to another. Finally, (A1) and (T8) taken together express the relational character of becoming of things: a thing may become with respect to one thing and not become with respect to another. This relational character of the notions A, Z, and B is a universal property since these pairs of theorems concern all things.

It is worthwhile devoting some attention to the relationship between the discussed notions A, Z, D, B and the notion of existence. Four different assumptions concerning the relationship between existence and past, present, and future have so far been discussed within PNF theory (cf. Chapter 5). For the issue under discussion, only two of them, namely

(Z) and (Z′), will be relevant. The first states that $E_x = N_x$, i.e., that only present events exist, the second that $E_x = P_x \cup N_x \cup F_x$, i.e., that all events exist.

Let us apply the first assumption, (Z), to the definition of the relations *A*, *Z*, *D*, and *B*. The desired connections between these relations and existence are in this case the following:

$$(1) \quad A(a|b) \rightarrow \bigvee_{y \in b} \bigwedge_{x \in a} [x \notin E_y] \wedge \bigvee_{y' \in b} \bigvee_{x \in a} [x \in E_{y'}]. \qquad \{D1, Z\}$$

In words: if the thing *a* comes-to-be with respect to the thing *b*, then the thing *a* (all the events *x*) does not exist with respect to a certain event *y* from *b*, although some events from *a* exist with respect to a certain event *y′* from *b*.

$$(2) \quad Z(a|b) \rightarrow \bigvee_{z \in b} \bigwedge_{x \in a} [x \notin E_z] \wedge \bigvee_{z' \in b} \bigvee_{x \in a} [x \in E_{z'}]. \qquad \{D2, Z\}$$

In words: if the thing *a* passes away with respect to the thing *b*, then the thing *a* (all events *x*) does not exist with respect to a certain event *z* from *b*, although some events from *a* exist with respect to a certain event *z′* from *b*.

$$(3) \quad D(a|b) \rightarrow \bigvee_{y \in b} \bigvee_{x \in a} [x \in E_y]. \qquad \{D3, Z\}$$

In words: if the thing *a* lasts with respect to the thing *b*, then a certain event from *a* exists with respect to a certain event from *b*.

$$(4) \quad B(a|b) \rightarrow \bigvee_{y \in b} \bigwedge_{x \in a} [x \notin E_y] \wedge \bigvee_{z \in b} \bigwedge_{x \in a} [x \notin E_z] \wedge \bigvee_{y' \in b} \bigwedge_{x \in a} [x \in E_{y'}]. \qquad \{D4, Z\}$$

In words: if the thing *a* becomes with respect to the thing *b*, then the thing *a* does not exist with respect to a certain event from *b* and does not exist with respect to some other event from *b*, but some event from *a* exists with respect to a certain event from *b*.

The consequences of the assumption (Z) imply that the relations of coming-to-be, passing away, lasting, and becoming have an existential character. This means that when the thing *a* comes-to-be, passes away, etc. with respect to the thing *b*, then certain events from *a* exist with respect to only certain events from *b* (those in the present of which they occur), while they do not exist with respect to other events from *b*. Hence, these events from *b* are distinguished with respect to existence.

Let us apply the second of the two assumptions, i.e., (Z′), to the definition of the relations *A*, *Z*, *D*, *B*. According to (Z′), every two events exist with respect to each other (regardless of what time relation they stand in). This being so, there is only one possible conclusion: if the thing a stands in any one of these relations with the thing b, then every event from a exists with respect to every event from b. Therefore, according to the assumption (Z′), the relations of coming-to-be, passing away, lasting, and becoming do not have an existential character. The above-mentioned conclusion expresses this fact. It may also be formulated thus: it is not a distinguishing feature of any of the events from b that certain events from a exist only with respect to them, when a stands in one of the considered relations with b.

Since the assumption (Z′) has been preferred to the assumption (Z) (cf. Chapter 5), the above relations obviously do not have an existential character. The coming-to-be, passing away, lasting, and becoming of things with respect to other things consist only in the fact that things stand in certain time relations to other things. This last conclusion seems to be inconsistent with common intuitions concerning the coming-to-be, passing away, lasting, and becoming of things. It is usually felt that a thing begins to exist when it comes-to-be, and that it ceases to exist when it passes away. This belief obviously assumes that only present events exist, i.e., the assumption (Z). This, however, is open to too many objections to be acceptable.

The following problem arises at this point: can the coming-to-be, passing away, lasting, and becoming of events with respect to things be described in a similar way? This is quite an interesting problem, although the precise sense of the notions used in this case must be different. The answer seems to be affirmative. Indeed, when one modifies the definitions (D1)–(D4), definitions of the relations A, Z, D, and B can be obtained for events. The expressions: $A'(x|a)$, $Z'(x|a)$, $D'(x|a)$, and $B'(x|a)$ read respectively: x comes-to-be with respect to a, x passes away with respect to a, x lasts with respect to a, and x becomes with respect to a. Let us note that the set of all events **S** is a domain of all these relations, while their counterdomain is the set of all things **T**; the relations are therefore non-homogeneous. Here are the definitions of the above relations:

(D1′) $$A'(x|a) \equiv \bigvee_{y \in a} [x \in F_y] \wedge \bigvee_{v \in a} [x \in N_v].$$

In words: the event x comes-to-be with respect to the thing a iff x occurs in the future of a certain event y from a and in the present of a certain (different) event v from a.

(D2′) $$Z'(x|a) \equiv \bigvee_{z \in a} [x \in P_z] \wedge \bigvee_{v \in a} [x \in N_v].$$

In words: the event x passes away with respect to the thing a iff x occurs in the past of a certain event v from a and in the present of a certain (different) event v from a.

(D3′) $D'(x|a) \equiv \bigvee_{v \in a} [x \in N_v]$.

In words: the event x lasts with respect to the thing a iff x occurs in the present of a certain event from a.

(D4′) $B'(x|a) \equiv \bigvee_{y \in a} \bigvee_{v \in a} \bigvee_{z \in a} [x \in F_y \wedge x \in N_v \wedge x \in P_z]$.

In words: the event x becomes with respect to the thing a iff there exist three such events y, v, z from the thing a that x occurs in the future of y, in the present of v, and in the past of z. Let us note that in the face of (D1′), (D2′), (D3′), definition (D4′) is equivalent to the theorem: $B'(x|a) \equiv A'(x|a) \wedge Z'(x|a)$, which can be simplified to: $B' = A' \cap Z'$, i.e., x becomes with respect to a iff x comes-to-be, lasts, and passes away with respect to a. The relation B' can therefore be defined by means of the relation A' and Z'.

Let us define the connections of inclusion between the relations A', Z', D', and B'. They are isomorphic with the connections between the relations A, Z, D, and B:

(T1′) $A' \subset D'$, {D1′, D3′}

(T2′) $Z' \subset D'$, {D2′, D3′}

(T3′) $B' \subset A'$, {D4′, D1′}

(T4′) $B' \subset Z'$, {D4′, D2′}

(T5′) $B' \subset D'$. {D4′, D3′}

Hence, if x comes-to-be or passes away with respect to a, then x lasts with respect to a; while x becomes with respect to a, then—as was said earlier—it comes-to-be, lasts, and passes away with respect to a.

As regards the mutual relationship between A' and Z', we may note that four possible cases occur, namely: $A' \cap Z'$, $A' \cap \overline{Z}'$, $\overline{A}' \cap Z'$, $\overline{A}' \cap \overline{Z}'$. These product relations are of course mutually exclusive. Any pair of the kind 'event–thing' stands in just one of these product relations. This means that these four cover all possible cases. The relation A' and Z' are not connected by the relation of inclusion in any way. It should be added that $A' \cap Z'$ is equal to the relation B', while the other three exclude this relation.

One may now ask, as before, whether every event stands in the relation Q to a particular thing, where Q now denotes any of the relations A', Z', B'.

As before, affirmative answers to this schema question are assumed axiomatically on the basis of experience. Here too it is enough to adopt only one axiom for the relation B':

$$(A1') \quad \bigwedge_{x} \bigvee_{a} [B'(x|a)].$$

In words: every event x becomes with respect to a certain thing a. On the basis of (D4′) the axiom (A1′) implies the following theorems:

$$(T6') \quad \bigwedge_{x} \bigvee_{a} [A'(x|a)], \qquad \{D4', A1'\}$$

$$(T7') \quad \bigwedge_{x} \bigvee_{a} [Z'(x|a)]. \qquad \{D4', A1'\}$$

In words: every event x comes-to-be with respect to a certain thing a (T6′), and every event x passes away with respect to a certain thing a (T7′).

It is worth considering the question which is a substitution of the schema: does every event not stand in a relation Q to a certain thing, where Q also represents any of the relations A', Z' or B'? The affirmative answers can only be assumed axiomatically. Two axioms have to be adopted; one for the relation A' and another for the relation Z':

$$(A2') \quad \bigwedge_{x} \bigvee_{b} \sim [A'(x|b)],$$

$$(A3') \quad \bigwedge_{x} \bigvee_{b} \sim [Z'(x|a)].$$

In words: every event x does not come-to-be with respect to a certain thing b (A2′), and every event x does not pass away with respect to a certain thing b (A3′). By virtue of the definition (D4′), the axioms (A2′) and (A3′) imply the theorem:

$$(T8') \quad \bigwedge_{x} \bigvee_{b} \sim [B'(x|b)]. \qquad \{D4', A2', A3'\}$$

In words: every event x does not become with respect to a certain thing b.

The theorems (T6′) and (A2′) together express the relational character of the coming-to-be of events, i.e., the fact that the same events come-to-be with respect to particular things and do not come-to-be with respect to others. Furthermore, (T7′) and (A3′) taken together express the relational character of the passing away of events—the same events pass away with respect to particular things and do not pass away with respect to others. Finally, (A1′) and (T8′) taken together express the relational character of the becoming of events: the same events become with respect to particular things and do not become with respect to others. These properties regarding the notions A', Z', and B' are universal since they concern all events.

To conclude this analysis of the relations discussed here, let us say a few words about their connection with the relation of existence. In this case too the problem will be investigated only from the point of view of the two assumptions (Z) and (Z′). The connections between the relations in question and existence are—under the assumption (Z)—the following:

(1′) $A'(x|a) \rightarrow \bigvee_{y \in a} [x \notin E_y] \wedge \bigvee_{v \in a} [x \in E_v].$ {D1′, Z}

In words: if the event x comes-to-be with respect to a, then x does not exist with respect to a certain event y from a, but does exist with respect to some other event v from a.

(2′) $Z'(x|a) \rightarrow \bigvee_{z \in a} [x \notin E_z] \wedge \bigvee_{v \in a} [x \in E_v]$ {D2′, Z}

In words: if the event x passes away with respect to a, then x does not exist with respect to a particular event y from a, but does exist with respect to some other event v from a.

(3′) $D'(x|a) \rightarrow \bigvee_{v \in a} [x \in E_v].$ {D3′, Z}

In words: if the event x lasts with respect to the thing a, then x exists with respect to a certain event v from a.

(4′) $B'(x|a) \rightarrow \bigvee_{y \in a} \bigvee_{v \in a} \bigvee_{z \in a} [x \notin E_y \wedge x \in E_v \wedge x \notin E_z].$ {D4′, Z}

In words: if the event x becomes with respect to a certain thing a, then x does not exist with respect to certain two events from a (y and z), but does exist with respect to a particular event from a (v), differing from them.

Accordingly, the relations of the coming-to-be, passing away, lasting, and becoming of events possess an existential character on the grounds of the assumption (Z). This means that if the event x comes-to-be, passes away, etc. with respect to the thing a, then it exists only with respect to certain events from a, and does not exist with respect to some other events from a. The former events (from a) therefore gain a privileged position.

Assuming (Z′), the connections between existence and the relations discussed here can be reduced to a common denominator and expressed in the following way: if the event x stands in any of these relations to the thing a, then the event x exists with respect to every event from a, since—as we know—according to (Z′), any two events exist with respect to each other. Hence, under the assumption (Z′), the relations A', Z', D', and B' do not have an existential character. In fact, in this case it is not a distinguishing feature of any event from a that the event x exists only with respect to it. In view of our preference for the assumption (Z′), it must be accepted that these relations are not existential.

It will now be demonstrated that the coming-to-be, passing away, lasting, and becoming of events can be discussed not only with regard to things, but also with regard to events. The new relations to be denoted by the symbols A'', Z'', D'', and B'', are generalizations of the corresponding relations A', Z', D', B'; this problem will be discussed in detail later on. The relations now introduced are homogeneous, i.e., their domains and counterdomains are the same set, namely the set S of all events—more simply, they are defined on the set S. However, they will differ by the number of places; A'' and Z'' have three, D'' is a binary relation, and B'' has as many as four. Let us proceed to a detailed consideration of these relations.

The expressions $A''(x|y, v)$ and $Z''(x|z, v)$ read respectively: x comes-to-be with respect to the pair y, v, and x passes away with respect to the pair z‘ v. The expression $D''(x|v)$ reads: x lasts with respect to v. Finally, the expression $B''(x|y, v, z)$ reads: x becomes with respect to the triple of events y, v, z. The definitions of these relations are as follows:

(D1″) $A''(x|y, v) \equiv [x \in F_y \wedge x \in N_v]$.

In words: the event x comes-to-be with respect to the pair of events y, v iff x occurs in the future of y and in the present of v.

(D2″) $Z''(x|z, v) \equiv [x \in P_z \wedge x \in N_v]$.

In words: the event x passes away with respect to the pair of events z, v iff x occurs in the past of z and in the present of v.

(D3″) $D''(x|v) \equiv [x \in N_z]$.

In words: the event x lasts with respect to the event v iff x occurs in the present of v. This relation of relative lasting corresponds to the relation of quasi-simultaneity.

(D4″) $B''(x|y, v, z) \equiv [x \in F_y \wedge x \in N_v \wedge x \in P_z]$.

In words: the event x becomes with respect to the triple of events y, v, z iff x occurs in the future of y, the present of v, and the past of z. The definition (D4″) is equivalent, by virtue of (D1″), (D2″), and (D3″), to the theorem $B''(x|y, v, z) \equiv A''(x|y, v) \wedge Z''(x|z, v)$. The theorem can be formulated more simply as $B'' = A'' \cap Z''$, i.e., x becomes with respect to the triple y, v, z iff x comes-to-be with respect to the pair y, v and passes away with respect to the pair z, v. The relation B'' can therefore be defined by means of the relations A'' and Z''. Let us note that in the framework of PNF theory, (D4$''$) is a reconstruction of the notion of becoming of events introduced by McTaggart (1927). This problem will be considered in detail in Chapter 7.

When one compares the definitions of the relations A', Z', D', and B' with the definitions of the relations A'', Z'', D'', and B'', the following connections between them become apparent.

If the event x comes-to-be with respect to the thing a, i.e., with respect to the events y and v from a, then it comes-to-be with respect to the pair of y and v. The inverse theorem of course does not hold, since the event x may come-to-be with respect to y and v, which do not belong to the same thing. Everything that has been said about the connection between A' and A'' obviously concerns the remaining pairs of relations Z' and Z'', D' and D'', B' and B''. These connections express the fact that the relations A'', Z'', D'', and B'' are in some sense generalizations of the respective corresponding relations A', Z', D', and B'. The former are obtained from the latter by abstracting from things and taking only events into account.

The connections between these relations can easily be established. If x comes-to-be or passes away with respect to y and v, then x lasts with respect to v; if x becomes with respect to y, v, and z, then it comes-to-be with respect to y and v, passes away with respect to z and v, and lasts with respect to v.

The question of whether every event stands in the relation Q to certain events—where Q denotes any one of the relations A'', Z'', D'', and B''—can also be raised in this context. This time, affirmative answers to the substitutions of this schema question can easily be proved within PNF theory (cf. Chapter 3). Therefore, they need not be axiomatically assumed, as was the case in the two previous series of relations. Here we have

(T6$''$) $\bigwedge_{x} \bigvee_{y,v} [A''(x|y, v)].$ {T27, T26, D1$''$}

In words: every event comes-to-be with respect to a certain pair of events.

(T7$''$) $\bigwedge_{x} \bigvee_{z,v} [Z''(x|z, v)].$ {T25, T26, D2$''$}

In words: every event passes away with respect to a particular pair of events.

(T8$''$) $\bigwedge_{x} \bigvee_{v} [D''(x|v)].$ {T26, D3$''$}

In words: every event lasts with respect to a particular event.

(T9″) $\bigwedge_{x} \bigvee_{y,v,z} [B''(x, |y, v, z)].$ {T25, T26, T27, D4″}

In words: every event becomes with respect to a particular triple of events.

The question (substitution of the schema) does every event not stand in a relation Q with certain events will be considered too. As before, Q denotes any of the relations A'', Z'', D'', B''. Affirmative answers can also be proved within PNF theory. Hence we have the following theorems:

(T10″) $\bigwedge_{x} \bigvee_{y',v'} \sim [A''(x|y', v')].$ {T30, T29, D1″}

In words: every event does not come-to-be with respect to a particular pair of events.

(T11″) $\bigwedge_{x} \bigvee_{z',v'} \sim [Z''(x|z', v')].$ {T28, T29, D2″}

In words: every event does not pass away with respect to a particular pair of events.

(T12″) $\bigwedge_{x} \bigvee_{v'} \sim [D''(x|v')].$ {T29, D3″}

In words: every event does not last with respect to a particular event.

(T13″) $\bigwedge_{x} \bigvee_{y',v',z'} \sim [B''(x|y', v', z')].$ {T30, T29, T28, D4″}

In words: every event does not become with respect to a particular triple of events. This version of the concept of becoming is discussed in Augustynek (1976a).

The two groups of theorems proved above lead to the following expected conclusions. The theses (T6″) and (T10″) taken together express the relational character of the coming-to be of events, i.e., the fact that the events come-to-be with respect to particular pairs of events and do not come-to-be with respect to other pairs. The theses (T7″) and (T11″) taken together express the relational character of the passing away of events—the events pass away with respect to a particular pair of events and do not pass away with respect to another pair. The theses (T8″) and (T12″) taken together express the relational character of the lasting of events—events last with respect to particular events and do not last with respect to others. Finally, the theses (T9″) and (T13″) taken together express the relational character of the becoming of events—events become with respect to particular triples of events and do not become with respect to other triples. The above properties of these relations are universal.

Let us consider the connections between existence and the relations A'', Z'', D'', and B''. As before, only the assumptions (Z) and (Z′) will be taken into account. Under the assumption (Z), these connections are as follows:

$$(1'') \quad A''(x|y, v) \rightarrow [x \notin E_y \wedge x \in E_v]. \qquad \{D1'', Z\}$$

In words: if the event x comes-to-be with respect to the pair of events y and v, then x does not exist with respect to y, but does exist with respect to v.

$$(2'') \quad Z''(x|z, v) \rightarrow [x \notin E_z \wedge x \in E_v]. \qquad \{D2'', Z\}$$

In words: if the event x passes away with respect to the pair of events z and v, then x does not exist with respect to z, but does exist with respect to v.

$$(3'') \quad D''(x|v) \rightarrow [x \in E_v]. \qquad \{D3'', Z\}$$

In words: if the event x lasts with respect to the event v, then it exists with respect to v.

$$(4'') \quad B''(x|y, v, z) \rightarrow [x \notin E_y \wedge x \in E_v \wedge x \notin E_z]. \qquad \{D4'', Z\}$$

In words: if the event x becomes with respect to the triple of events y, v, z, then it does not exist with respect to y and z, but does exist with respect to v.

The relations A'', Z'', D'', B'' therefore possess an existential character in the light of the assumption (Z). This means that if an event comes-to-be or passes away with respect to a particular pair of events, then it exists with respect to only one member of this pair, but not with respect to the other member. If the event becomes with respect to a particular triple of events, then too it only exists with respect to one member of the triple. These events therefore gain a privileged position from an existential point of view.

The connections between existence and the relations A'', Z'', D'', and B'' look different in the light of the more liberal assumption (Z′). They will not be separately formulated since they are quite similar and may be expressed in the following way: if the event x stands in one of these relations with another event (D''), a pair of events (A'', Z''), or a triple of events (B''), then x exists with respect to every one of these events. This stems from the fact that, according to the assumption (Z′), any two events exist with respect to each other. Hence, under this assumption, these relations do not have an existential character. Actually, it is not a distinguishing feature of any of the events to which x is related that x exists only with respect to this event. Since the assumption (Z′) is preferred in the present considerations, the relations A'', Z'', D'', and B'' are denied an existential character. These relations therefore only consist in the fact that events have certain temporal connections with particular other events precisely as the appropriate definitions of these relations describe.

Two general remarks are appropriate to conclude this chapter. First, the definitions of the three sets of relations, as well as the theorems concerning them formulated within PNF theory can easily be reiterated within the framework of standard $P^uN^uF^u$ theory. Secondly, I believe that the results obtained in this chapter, regardless of the closeness of the defined relations to their commonsense analogues, represent an important step in the further development of the eventism assumed within PNF theory, and especially of that component of eventism which explicitly contains the notion of the thing.

CHAPTER 7

OTHER THEORIES OF PAST, PRESENT, AND FUTURE

To conclude this discussion, it is essential to include at least a brief analysis of some theories of past, present, and future that differ from the one presented here. There are at least two reasons for undertaking this analysis. To begin with, these theories are in sharp contrast to ours and differ from it very considerably. Secondly, they are widespread and have many supporters among representative groups of writers on the philosophy of time. These theories or, more precisely, their main assumptions, as well as some important implications of these assumptions, will be presented in this chapter. They will also be compared and contrasted with the theory presented in this work.

First we introduce two dichotomous classifications of the theories of past, present, and future. Both these classifications are vital to the present considerations.

The first is to establish whether past, present, and future are regarded as possessing a relational character or not, i.e., whether these sets may be said to be related through appropriate relations to certain objects; in other words, whether they are, or are not, defined by means of adequate time relations.

Any theory in which past, present, and future have a relational character will be called *relational*; otherwise, it will be called *absolute*. The very principle of this classification implies that it is dichotomous; a theory must be either relational or absolute.

The idea of the second classification is to establish whether past, present, and future are regarded as objective, i.e., subject-independent, or whether they are regarded as subjective, i.e., subject-dependent. In other words, the question is whether the classification of events (or moments) as past, present, or future is objective or subjective. If past, present, and future possess an objective character in a theory, it will be called *objectivistic*, otherwise it will be called *subjectivistic*. This classification again implies dichotomy: a theory must be either objectivistic or subjectivistic.

Let us now first compare and then combine these classifications. It is obvious that the first is ontological, while the second is distinctly epistemological; they are therefore very different from the philosophical point of view. Combining these classifications, we obtain four possible logical types of theories of past, present, and future as follows: (1) relational and objectivistic, (2) relational and subjectivistic, (3) absolute and objectivistic, (4) absolute and subjectivistic. It will be more rigorously demonstrated below that the theory put forward in the present book belongs to the first of these four types, i.e., that it is relational and objectivistic.

The first theory to be discussed in this chapter belongs to the second, i.e., relational and subjectivistic type, the other belongs to the third type, which is absolute and objectivistic.

No theory of the fourth type—absolute and subjectivistic—is considered here. I would suggest that the views of Kant would probably fit into this category. His subjectivism (more precisely, transcendental intersubjectivism) concerned not only past, present, and future, but also, and above all, time itself. This last conception seems to me particularly devoid of foundation—but that is another matter.

We shall begin with the presentation and criticism of the relational-subjectivistic theory since it is more comprehensible and, being relational, is closer to the relational-objectivistic theory developed in the present book.

Several versions of this theory have been put forward, but the most developed and widespread is the linguistic version formulated by Bertrand Russell (Russell, 1903) and supported by such philosophers as W. V. Quine (Quine, 1953) and N. Goodman (Goodman, 1951).

To reconstruct this theory, a few simple formalizations will be used, which will also allow for comparison with our theory. The following symbols will be involved: the set symbols P_x, N_x, F_x, the relation symbols W, R, $\breve{W}$ (these are understood classically), as well as three binary relation symbols—*UP*, the articulation of the statement that something is in the past; *UN*, the articulation of the statement that something is at present; *UF*, the articulation of the statement that something is in the future.

Thus the expressions of the form $UP(x, y)$, $UN(x, y)$, $UF(x, y)$ read respectively: the event x is an utterance of the statement that the event y is in the past; the event x is an utterance of the statement that the event

y is at present; the event x is an utterance of the statement that the event y is in the future.

It should be noted that the variables x and y are assumed to run over the set of (point-like) events; this assumption has a definitely idealizational character in the case of the variable x (the duration of the utterance is not taken into account).

The linguistic relational definitions of the sets P_x, N_x, F_x are as follows:

(1) $P_x \stackrel{\text{def}}{=} \{y\colon W(y,x) \wedge UP(x,y)\}$,

(2) $N_x \stackrel{\text{def}}{=} \{y\colon R(y,x) \wedge UN(x,y)\}$,

(3) $F_x \stackrel{\text{def}}{=} \{y\colon \breve{W}(y,x) \wedge UF(x,y)\}$.

In words: the past of the event x is the set of such events y that y is earlier than x, and x is the utterance of the statement that y is in the past; the present of the event x is the set of such events y that y is simultaneous with x, and x is the utterance of the statement that y is at present; the future of the event x is the set of such events y that y is later than x, and x is the utterance of the statement that y is in the future.

The origin of the theory presented here is to be found in the question of the actual meaning of the temporal statements: 'y is in the past', 'y is in the present', and 'y is in the future'. The answers to the question are just these definitions which in non-formalized form read respectively: (1) the statement 'the event y is in the past' means the same as the statement 'the event y is earlier than the utterance of the statement "the event y is in the past"'; (2) the statement 'the event y is at present' means the same as the statement 'the event y is simultaneous with the utterance of the statement "the event y is at present"'; (3) the statement 'the event y is in the future' means the same as the statement 'the event y is later that the utterance of the statement "the event y is in the future"'.

We can see that, according to these definitions, past P_x, present N_x, and future F_x refer to definite linguistic events which are the utterance (or writing down) of the corresponding sentences given above. Together with the given non-formalized definitions of P_x, N_x, F_x, this facilitates our understanding of this conceptually quite difficult theory.

The above definitions can be considerably simplified on the assumption that the following implications (empirically) hold: $UP(x,y) \rightarrow W(y,x)$, $UN(x,y) \rightarrow R(y,x)$, and $UF(x,y) \rightarrow \breve{W}(y,x)$. (The sentences $UP(x,y)$, $UN(x,y)$, and $UF(x,y)$ are of course assumed to be true.) The following definitions are then obtained by means of tautology: $[(p \rightarrow q) \rightarrow (p \wedge q \equiv p)]$.

These definitions are structurally identical with their non-formalized counterparts:

(1′) $P_x \overset{\text{def}}{=} \{y\colon UP(x, y)\}$,

(2′) $N_x \overset{\text{def}}{=} \{y\colon UN(x, y)\}$,

(3′) $F_x \overset{\text{def}}{=} \{y\colon UF(x, y)\}$.

The linguistic theory enables one to formulate a definition of the becoming of events with respect to triples of events, as in Chapter 6. By virtue of the simplified linguistic definitions of the sets P_x, N_x, F_x, the following formula is obtained:

$$B(x|y, v, z) \equiv [UF(y, x) \wedge UN(v, x) \wedge UP(z, x)].$$

In words: the event x becomes with respect to the triple of events y, v, z iff y is the utterance of the statement that x is in the future, v is the utterance of the statement that x is at present, and, finally, z is the utterance of the statement that x is in the past. This description provides the linguistic version of the notion of becoming of events.

The linguistic versions of the definitions of the coming-to-be and passing away of events with respect to pairs of events can be formulated in a similar way on the basis of the definitions of these notions in Chapter 6. By means of the simplified linguistic definitions of the sets P_x, N_x, F_x, the following formula is obtained:

$$A(x|y, v) \equiv [UF(y, x) \wedge UN(v, x)].$$

In words: the event x comes-to-be with respect to the pair of events y and v iff y is the utterance of the statement that x is in the future and v is the utterance of the statement that x is at the present.

$$Z(x|v, z) \equiv UN(v, x) \wedge UP(z, x).$$

In words: the event x passes away with respect to the pair of events v and z iff v is the utterance of the statement that x is at the present and z is the utterance of the statement that x is in the past.

It is now appropriate to compare the linguistic version of the relational-subjectivistic theory of past, present, and future described here with the theory of these objects presented in this book. This enables us to criticize the linguistic theory and demonstrate the comparative advantages of our theory.

The accepted definitions of past, present, and future should be recalled in order to reach these two objectives:

$$P_x \stackrel{\text{def}}{=} \{y\colon W(y, x)\},$$

$$N_x \stackrel{\text{def}}{=} \{y\colon R(y, x)\},$$

$$F_x \stackrel{\text{def}}{=} \{y\colon \breve{W}(y, x)\}.$$

To facilitate the argument, it will be assumed that the time relations W, R, $\breve{W}$—earlier, simultaneously, and later—have their classical meaning (i.e., are absolute relations within the framework of classical mechanics) just as in the above linguistic version of the relational-subjectivistic theory.

A few important conclusions can be reached by a comparison of these definitions with the linguistic definitions of the sets P_x, N_x, F_x. In the first place, let us examine the issue from a purely formal point of view. First, the linguistic definitions determine the sets P_x, N_x, F_x in the same relational way as our definitions do. The sets are determined by means of the time relations $W, R, \breve{W}$. In other words, these sets are always related to certain events. Therefore, the theory of which a version is considered here is regarded as relational.

Secondly, the linguistic definitions comprise certain additional conditions which are imposed upon the events standing in the time relations W, R, $\breve{W}$. These conditions are expressed in the linguistic definitions by the relational predicates *UP*, *UN*, *UF*. They essentially limit (restrict) the sets P_x, N_x, F_x, and sometimes even reduce them to unit sets. (In the case of a fixed event x, e.g., an utterance that y is in the past, P_x—the past of x—is a unit set containing only the event y, since the utterance of the statement that a certain event is in the past corresponds to this single event mentioned in the statement.) Of course, the conditions limit these sets when compared with the sets P_x, N_x, F_x defined (as above) only by means of $W, R, \breve{W}$.

Let us note that conditions limiting these sets may be other than linguistic. One can imagine conditions of a causal nature. For example, the past of the event x can be defined by means of a causal relation in the following way: $P_x \stackrel{\text{def}}{=} \{y\colon W(y, x) \wedge H(y, x)\}$, analogously, the future of the event x: $F_x \stackrel{\text{def}}{=} \{y\colon \breve{W}(y, x) \wedge \breve{H}(y, x)\}$, and the present of the event x: $N_x \stackrel{\text{def}}{=} \{y\colon R(y, x)\}$. Consequently, the sets P_x, N_x, F_x cease to be singletons under the quite certain assumption that an event possesses many different causes and many different results.

How should the above formal conclusions be viewed? The first conclusion, which states the relational character of the theory, appears to be an important advantage. This view will be justified later, when the absolute theory of past, present, and future is presented and criticized.

The second conclusion, concerning the above-noted limitations of the sets P_x, N_x, F_x by the corresponding definitions, appears to be a substantial weakness of this, as well as any other theory (for instance, of those *ad hoc* formulated causal definitions of the sets P_x, N_x, F_x). In my view, there is no sensible empirical, linguistic (meaning) or logical reason to support such a limitation.

A very important point should now be stressed; namely, the limiting conditions involved in the definitions examined here concern relations of an overtly subjective character, i.e., relations which at least have domains consisting of subject-dependent events. In fact, the domains of the relations *UP*, *UN*, *UF* are the sets of utterances of statements concerning the past, present, and future of events. The sets P_x, N_x, F_x are therefore explicitly related to certain linguistic events.

Past, present, and future P_x, N_x, F_x consequently possess a subjective character within linguistic theory. Hence the theory is classified as subjectivistic. If, therefore, nobody utters a statement claiming that y is in the past, then—according to the linguistic theory of the past—y does not belong to the past of x despite the fact that y is earlier than the event x; formally: $\sim UP(x, y) \rightarrow \sim y \in P_x$. Similar implications concern events about which nobody utters statements claiming that they are present or future: they do not then belong either to the present or to the future.

In the examples discussed so far, we have assumed that the subject (or language user) performed no activity regarding particular events, i.e., he did not utter any statements about the past, present, or future of these events. However, much more drastic circumstances can be imagined, namely, a world which does not contain any language users at all.

It is certain that in such an imaginary world—which is usually presupposed, in order to enter into polemics with subjectivist idealism—no event can, according to the linguistic definitions, be past, present, or future (with respect to some other events, of course). I do not believe, however, that we need a fictious world deprived sentient beings for this purpose. The existing world, in which some events are outside the scope of human linguistic activity, is quite sufficient.

It is not difficult to establish the likely consequences of the subjectivization of past, present, and future for the notions thus defined, such as becoming, as well as coming-to-be and passing away. Clearly, these notions turn out to be purely subjective too. If a certain individual claims something about a certain event (e.g. about the eclipse of the moon), for instance, that it is in the future, further, that it is taking place at present, and, finally, that it took place in the past, then, according to the linguistic definition, the event becomes with respect to this individual. If, however, even a single one of these conditions is not satisfied, then, according to the same definition, the event does not 'become'. In this way, the becoming of events as well as their coming-to-be and passing away have a clearly subjective character, depending on the appropriate activity of the knowing subject.

Supporters of the so-called absolute-objectivistic theory (cf. Gale, 1968), who identify the becoming of events with the passage of time, think that this theory recognizes this passage as subjective, treating it as a linguistic, and therefore subjective phenomenon.

I should make it clear that I utterly disagree with the conclusions described above. The subjectivization of the notions of past, present, and future and the concepts that are definitionally derivative with resepect to them (becoming, coming-to-be, passing away) is totally unacceptable. In my opinion, neither physics nor any other natural sciences or humanities are concerned with subjective notions of past, present, and future; nor are they understood in this way in everyday life. Moreover, they are never used in the linguistic sense in science or in everyday life.

It becomes clear after closer investigation that in science, and particularly in physics, the notions of past, present, and future, and the terms definitionally derivative with respect to them, are understood realistically, i.e., objectively; they are also applied in this sense. If these notions are defined by means of the time relations earlier, simultaneously, and later, no other conditions are imposed on them — I mean of course conditions of an objective character, e.g. the above-mentioned causal conditions which could possibly be (alternatively) applied. Thus, the notions of past, present, and future acquire a universal character.

A question arises at this point—how to reconcile the subjective character of the linguistic version of the theory with the undoubted ontological realism represented (in time theory as well) by its founder (Russell) and other supporters (Quine, Goodman, etc.)? Since the linguistic theory cannot be dismissed as a purely nonsensical brainchild of Russell and its other supporters, another question arises—if this approach can be logically consistent, what are the actual arguments underpinning linguistic theory?

As far as the problem of the consistency of linguistic theory is concerned, the relations W, R, $\breve{W}$, as well as time, understood as some determined structure of the objectively existing world, can be regarded as objective. At the same time, past, present, and future can be claimed to be subjective, i.e., subject-dependent. This stems from the fact that the assumption of the objectivity of the appropriate relations and of the objectivity of time does not (directly) imply the thesis of the objectivity of past, present, and future. The implication holds provided that these objects are defined only by means of relations (or properties) which are regarded as objective. This is exactly the case in our relational-objectivistic theory where the sets P_x, N_x, F_x are defined by means of the relations W, R, $\breve{W}$ only. Therefore, there is no discrepancy from the logical point of view.

As for the problem of argumentation is concerned, it has already been mentioned that the origin of the theory must be sought in the question of the actual meaning of the temporal statements: 'x is in the past', 'x is at present' and 'x is in the future'. Proceeding from this, we arrive by semantic analysis at the linguistic definitions of these statements given above. Particular arguments to support the theory will not be presented here. Criticism will be confined to the inadequacy of linguistically defined notions of past, present, and future in science and everyday life that was objected to earlier.

Another problem cannot be overlooked in this context. It is undoubtedly the case that if x is an utterance of the statement that y is past, then y is in the past with respect to x; analogously, if x is an utterance of the statement that y is at present, then y is at present with respect to x; and if x is an utterance of the statement that y is in the future, then y is in the future with respect to x. The truth of the statements 'x is in the past', 'x is at present' and 'x is in the future' is of course assumed at this point.

In other words, the following implications should be accepted:

(i) $UP(x, y) \rightarrow y \in P_x$,

(ii) $UN(x, y) \rightarrow y \in N_x$

(iii) $UF(x, y) \rightarrow y \in F_x$.

It is not a universal truth, however, that if y is in the past with respect to x, then x is an utterance of the statement that y is in the past, neither it is true that if y is at present, then x is an utterance of the statement that y is at present; nor, finally, is it true that if y is in the future with respect to x, then x is an utterance of the statement that y is in the future. In other words, implications inverse to those given above are false when regarded as universal theorems.

I therefore recognize the antecedents of the formulae (i), (ii), (iii) to be effective linguistic criteria (sufficient conditions) for accepting the truth of their consequences, i.e., statements concerning the relative past, present, and future of events. These are, of course, very special criteria. From the point of view of our theory, the general criteria for accepting the statements $y \in P_x, y \in N_x, y \in F_x$ are, obviously, the acceptances of the statements $W(y, x), R(y, x), \breve{W}(y, x)$, respectively. This follows from the definitions of the notions of P_x, N_x, F_x.

Where Russell and other supporters of linguistic theory presumably went wrong was in adopting the (very special) criteria of past, present, and future as their definientia, thus confusing them with definitions. This is probably one of the origins of their theory.

Let us now proceed to the presentation and criticism of the theory referred to at the beginning as absolute-objectivistic. Its supporters are to be found mainly among analytical philosophers of time, such as P.F. Strawson (Strawson, 1952) and W. S. Sellars (Sellars, 1962) among others. Its leading representative is R. Gale, the author of a comprehensive study of the language of time (Gale, 1968). It is this work of deep insight, which will mainly be referred to in the following remarks.

The central thesis of this theory is the statement claiming that past, present, and future are not related to particular events (or some other objects), but possess a non-relational (in this sense, absolute) character. In the present formulation of the theory, we treat past, present, and future as sets of events. Accordingly, they are denoted by P, N, F, without the event-index.

In other words, this thesis claims that the sets P, N, F are not definable by means of any particular relations, in particular, they are not definable by means of the time relations W, R, $\breve{W}$. The expressions $x \in P, x \in N$, $x \in F$ are used here as expressions *sui generis*. According to Gale, an event is simply past, present, or future. This conception is supposed to be underpinned by colloquial speech understanding, an understanding for which much is claimed—it was also supposed to support linguistic theory!

This thesis is supplemented by the additional claim that the time relations $W, R, \breve{W}$ not only are but must be defined by means of the absolute sets P, N, F. To justify this view, it is asserted that these relations have a time-character only because their constituents (i.e., events) possess the characteristics of P, N, F which are fundamental from the point of view of the 'essence' of time.

The second key thesis of the theory claims that this essence of time consists in the fact that something known as time-becoming is characteristic of all events. This 'becoming' is the change which events undergo with respect to the characteristics of P, N, and F, and, in fact, means the passing of events from future F, through the present N to the past P. According to some authors, this direction of passing is the direction of time. Let us note that the time-becoming is often called the passage of time, as distinct from the change in time (e.g. of a thing). Since P, N, F are non-relational, becoming is also non-relational and absolute.

The assumption accepted here that only present events exist is also obviously construed absolutely as $E = N$. In consequence, the becoming of events has an explicitly existential character. This means that an event enters into existence when passing from the future F to the present N, and leaves existence when passing from the present N to the past P.

According to supporters of this theory, only when interpreted thus, do past, present, and future, and, consequently, time-becoming have an objective character and are not subject-dependent. This thesis follows from the tacit assumption that a relational conception of these characteristics can only be subjective. The linguistic theory discussed above is therefore treated as being subjective.

Let us analyse these theses and try to find out, roughly at least, how to actually understand this theory which so far does not seem particularly easy to comprehend. The main difficulty lies in the lack of relativization of the sets P, N, F, i.e., in their absolute character, postulated as the theory's basis. In other words, the difficulty resides in our understanding of the meaning of the expressions $x \in P$, $x \in N$, $x \in F$. The first interpretation which suggests itself is the following: there is only one partition of the set of all events $\mathbf{S}$ into the sets P, N, F, i.e., only one past P, one present N, and one future F.

This supposition, however, is essentially inconsistent with other theses of the theory since it implies—by virtue of the fact that the sets P, N, F are assumed to be disjoint—that each event is permanently attached to one of them. However, when speaking about the same event, we often refer to it as past, present, and future, without being inconsistent, since we have in mind different moments of time. Colloquial speech—the alleged basis of this theory—therefore contradicts this implication in an obvious manner.

Secondly, the other thesis of the theory—concerning absolute time-becoming—claims that each event passes from the future F, through the present N to the past P, which again contradicts the above implication, and therefore also the interpretation stipulating the unique partition of the set $\mathbf{S}$ into the sets P, N, F. Hence, we are forced to relinquish this interpretation—as do the majority of its adherents, including Gale.

As a result, some of the supporters of the theory assume that the present N, which cuts the set $\mathbf{S}$ into the sets P and F, shifts, resulting in the consecutive and obviously different partitions of the set S into the sets P and F. Gale regards the notion of shifting 'Now' to be a metaphor and the whole idea to be senseless, since he argues that this would be tantamount to reification of the present (and past and future as well) and therefore to recognition of it as a unique object. It is amazingly inconsistent, however, that at the same time Gale does not claim that different presents exist.

Independently of these internal controversies, the supporters of the theory under discussion are simply forced by the reasons given above to accept (though only implicitly) many different presents and also many different pasts and futures. However, they seem to be unaware of the consequences; namely, that the above assumption necessarily implies that these pasts, presents, and futures, just because they are different, must be related to something—events or at least moments. This being the case, the main thesis of the theory cannot in my opinion hold. In other words, the sets P, N, F cannot be treated in any other than a relational way.

It is worthwhile dwelling on the reasons for the stubborn persistence of the absolute point of view criticized here. The main reason would seem to be the fact that, probably due to economy of words, colloquial speech uses such shorthand expressions concerning past, present, and future as $x \in P$, $x \in N$, $x \in F$, without reference to particular events. This usage has been absolutized by certain philosophers and accepted as corresponding to the real state of affairs. In actual fact what we have here are merely abbreviations of complete temporal expressions such as $x \in P_y$, $x \in N_y$, $x \in F_y$. These expressions explicitly contain the references of the sets P, N, F to certain events. The shorthand expressions are always used with a tacit assumption concerning the appropriate reference of the sets P, N, F. Abbreviations without these assumptions understood literally (not as abbreviations) are obviously devoid of any meaning. The indicated source of the fallacy of the theory under discussion seems to be the strongest argument against it.

The problem of the definitional connections of past, present, and future and the time relations earlier, simultaneously, and later will be considered next. This problem is related to the first thesis of the theory under discussion. The theory claims firstly that the former cannot be defined by means of the latter, i.e., that this cannot be done without loosing the actual meaning of the notions of past, present, and future. This is a sound argument provided that the sets P, N, F are treated in a non-relational way, although then there is no way of knowing what the argument is about, since such definitions cannot be constructed for purely formal reasons.

Yet, this is by no means the case if the sets P, N, F are treated in a relational way. The theory of past, present, and future which has been developed in the present book on the basis of the relations earlier, simultaneously, and later is the best evidence. Let us add that nothing is lost when the sets P_x, N_x, F_x are defined by means of the relations W, R, $\breve{W}$; in particular, the property of objectivity of the sets P, N, F is not lost since the relations have an objective character. It should be noted that only the symbols W, R, $\breve{W}$ appear in the definientia of the definitions of P_x, N_x, F_x in this case (there are no additional predicates, e.g. linguistic predicates).

Secondly, the theory under discussion states that the relations earlier, simultaneously, and later not only can, but also should be defined by means of past, present, and future, i.e., reduced through definitions to the latter, regarded as fundamental concepts. Now, it has never been possible to carry out this stipulation given the assumption that P, N, F are non-relational, in spite of great efforts on the part of supporters of the theory.

In his book Gale presents as many as seven attempts at definitions of this kind (the majority are his own). Nevertheless, he comes to the bitter conclusion that even the most promising of them (which are exceedingly complicated) contain *circulus in definiendo*. This is so because W, R, $\breve{W}$ persistently appear in the definientia of their own definitions (by means of P, N, F)! This is actually an open recognition of the important weakness in this part of the theory. In my view, the problem of the reduction of the relations W, R, $\breve{W}$ to the sets P, N, F is meaningless, since a non-relational approach to these sets does not make sense.

There is some point in asking, however, whether the time relations W, R, $\breve{W}$ can be defined by means of the relationally treated sets P_x, N_x, F_x within our relational-objectivistic theory. This is certainly possible through 'reversing' the definitions (D2), (D3), (D4) from Chapter 3. Here is the result within PNF theory:

$$W(y, x) \overset{\text{def}}{=} y \in P_x,$$

$$R(y, x) \overset{\text{def}}{=} y \in N_x,$$

$$\breve{W}(y, x) \overset{\text{def}}{=} y \in F_x.$$

In words: x is earlier than y iff y belongs to the past of x; x is quasi-simultaneous with y iff y belongs to the quasi-present of x; x is later than y iff y belongs to the future of x.

The problem is, however, whether there are any syntactical (or maybe also semantic) reasons making it more appropriate and convenient to treat the relations W, R, $\breve{W}$ as definitionally derivative with respect to the sets P_x, N_x, F_x (or better still, with respect to the relations of containment of the events in sets, i.e., relations $y \in P_x$, $y \in N_x$, $y \in F_x$) than to treat them (or, more precisely, the relation W only) as primitive notion of the theory.

It is now time to consider the second key thesis of the theory considered here which claims that the essence of time consists in the becoming of events, i.e., in their passing from the future, through the present, to the past. Becoming is usually identified with the passage of events, or directly with a change of time. The time-becoming of events obviously possesses an absolute character since it is definitionally derivative from the absolute P, N, F. Since non-relational notions of P, N, F are meaningless, as we attempted to demonstrate, the non-relational notion of becoming is also meaningless. The same is true of the coming-to-be and passing away of events defined analogously, i.e., by means of P, N, F.

As we saw in Chapter 6, the relational notions of becoming, coming-to-be, and passing away can be successfully, and therefore meaningfully, defined in a relational way (by means of P_x, N_x, F_x). Moreover, these notions apply not only to events, but also—and perhaps above all—to things. This example shows how much we can gain by being ready to change (even slightly) our conceptual framework.

The question now arises of how to understand this 'passage' or 'change' of time. Gale thinks that these are only metaphors, poetical names for the time-becoming of events, which in the framework of this conception is understood absolutely. If that is so, the notion of passage of time is equally metaphorical, as is the notion of becoming (as was pointed out earlier). Gale's position stems from his awareness of the fact that a literal interpretation of the notion of passage or change of time raises at least two troublesome questions which are traps for the unwary.

Firstly, is there any sense in the notions of passage or change of time? If there is, then one must accept that time changes in time, i.e., in itself, which is nonsensical. To avoid this trap consequence, one must assume some other time, a supertime (or meta-time) within which our common time flows or changes. Now, however, an analogous question concerning change can be put with respect to this supertime, etc. This way we reach a *regressus ad infinitum*. To sum up: we end up with either nonsense or infinite regression.

Secondly, if time flows or changes, another question arises: how fast does it do so? It is clear that no answer to this question can be meaningful.

It is not difficult to note that the notion of passage or change of time when taken literally (which gives rise to both these questions) tacitly assumes the reification of time. Time is treated like other objects which undergo changes that proceed at a certain speed.

Cannot the notion of passage of time be rationalized while at the same time preserving the common intuitive meaning? I think it can. It seems that when we speak about the passage of time what we have in mind is the passing of things (including persons) and events. Here, this notion appears to coincide with the meaning of the notions of the becoming of things (with respect to things) and the becoming of events (with respect to things or triples of events) introduced in the previous chapter. These notions, which were defined there only by means of the relational notions of past, present, and future, cannot easily be accused of lacking sense or diverging from common intuition. Even less can they be accused of involvement in pseudoproblems. Let us add that such an interpretation rehabilitates the notion of the passage of time (or change of time), which has been so strongly attacked by scientifically-oriented philosophers of time.

To end this chapter, and the book as a whole, it would seem worthwhile to give an overall evaluation of the theories outlined and criticized here, as well as of the author's own theory as presented in this work.

The basic weakness of the relational-subjectivistic theory, and, in particular, its linguistic version which has been considered here, is certainly the subjective nature it postulates for past, present, and future, and, as a consequence, also for other notions derived from them. This subjectivization is at odds with the use of these notions in science and in everyday life, where commonsense and colloquial speech are dominant.

However, one cannot overlook the undoubted merits of this theory. The relational characterization of past, present, and future is certainly correct, as is the use of the relations earlier, simultaneously, and later in order to effect this characterization. The criticism of the absolute interpretation of these notions typical of absolute theories is also correct. This very subtle and effective criticism has frequently been helpful in the present confrontation with the absolute-objectivistic theory. Finally, the formulation of some (linguistic) criteria of the notions of past, present, anf future has been an important contribution.

The basic weakness of the absolute-objectivistic theory is the attempt to provide an absolute characterization of past, present, and future. I call this an 'attempt' since, as has been demonstrated, this characteristic is meaningless even from the point of view of colloquial speech from which it claims to be derived. As a result, such notions as absolute becoming, which is defined by means of those notions mentioned above, are also meaningless. The status of this theory is in my opinion much inferior to that of the relational-objectivistic theory, since it is a more significant violation of the 'rules of the game' to make nonsensical statements than to make false statements.

It would, however, be unjust to deny that this theory has any merits. In spite of its attempt at absolutization, the persistent maintenance of an objectivistic characteristic of past, present, and future, and such derivative notions as the becoming of events deserves approval. The second advantage is the uncompromizing defence of this objectivistic point of view against subjective theories.

I have greatly benefited from this critical defence; I refer here in particular to the arguments presented in the frequently cited book by R. Gale. I can only regret that his criticism is also aimed against the relational characteristic of past, present, and future. In addition, this criticism makes the tacit and entirely false assumption that a relational approach to these notions implies their subjectivization.

The theory put forward in the present book is relational as well as objectivistic, and does not have the weaknesses of either of the other theories evaluated above. Firstly, it characterizes past, present, and future in a relational way by means (only) of the appropriate relations earlier, simultaneously, and later. It thus avoids all objections raised against the absolute-objectivistic theory. Secondly, by using relational categories and postulating the objectivity of time relations, this method implies that past, present, and future are treated objectively. (*Nota bene*: the postulate of the objectivity of relations is explicitly accepted by supporters of both critized theories.) In this way, the theory avoids arguments raised against the relational-subjectivistic theory.

Does this mean that the proposed theory is perfect? It certainly does not—although it is better than either of its competitors with respect to the two important properties mentioned above. Outside the framework presented there is certainly still room for change, as new developments occur in physics and philosophical reflection on time.

APPENDIX

POINT-EVENTISM

This appendix presents an outline of an ontology that might be called *point-eventism*. This has been suggested by H. Reichenbach (Reichenbach, 1924 and 1928) and R. Carnap (Carnap, 1929), and then developed by H. Mehlberg (cf. Mehlberg, 1980). It seems to me that this ontology accords better than any other with both the General and the Special Theory of Relativity.

While understanding the term 'event' to mean point-event, the main thesis of point-eventism states that every object is an event or a set founded in events. The last expression means that it is a set of events or a family of some sets of events, and so on.

This thesis presupposes that only events are individuals (i.e., non-sets); all other objects are sets founded in events. The term 'set' is used here in the set-theoretical sense, which implies the acceptance of an ontological commitment to the existence of sets founded in events (or at least some of them).

I treat events as objects which are not extended in time and space (in the everyday, non-defined sense). As such, they are exactly space-time localizable. That is why we speak here of point-events, and point-eventism. This treatment of events is the usual rule in relativity, and sometimes appears in natural language.

The set of all (point) events I denote by S, and its elements by x, y, z, etc. The set S has, according to the Special Theory of Relativity (STR), two types of relations. The first type includes *absolute relations*, i.e., those independent of any arbitrary inertial reference system. Some time and space relations and also causal relation belong here. The second type includes *relative relations*, i.e., those dependent on an inertial reference system. Only certain time and space relations belong to this type.

Absolute relations yield eventistic definitions of such categories of objects as things, processes, cross-sections, and so-called *coincidents*. Relative relations yield eventistic definitions of such kinds of objects as moments, space points, and space-time points. These definitions substantiate the main thesis of point-eventism formulated at the beginning.

Absolute time relations are: (1) *earlier than* (briefly, *earlier*) W and *later than* (briefly, *later*) $\breve{W}$; both are asymmetric and transitive (ergo irreflexive) in S; (2) *quasi-simultaneous* $\overline{R}$, where $R = \overline{W} \cap \overline{\breve{W}}$; it is symmetric and reflexive but not transitive in S, therefore it is a relation of time similarity only; (3) *time separation* $\overline{R}$, where $\overline{R} = W \cup \breve{W}$; it is symmetric and irreflexive but not transitive in S, ergo $\overline{R}$ is a relation of time difference.

The relations W and R have the following light-conical interpretations. The events x and y enter into W, i.e., Wxy takes place, iff x lies inside or on the surface of the lower light cone of y. The events x and y enter into R, i.e., Rxy takes place, iff x lies outside the light cone of y or is identical with y. Of course, the light-conical interpretations of the relations $\breve{W}$ and $\overline{R}$ follow logically from the above interpretations of W and R.

Absolute space relations are: (1) *space separation* $\overline{L}$, which is symmetric and irreflexive in S, ergo $\overline{L}$ is a relation of space difference; (2) *quasi-co-location* L, which is symmetric and reflexive but not transitive in S, hence it is the relation of space similarity only. Here the light-conical interpretations are as follows. The events x and y enter into $\overline{L}$, i.e., $\overline{L}xy$ takes place, iff x lies outside or on the surface of light cone of y. The events x and y enter into L, i.e., Lxy takes place, iff x lies inside the light cone of y or is identical with y.

Only one relation belongs to absolute space-time relations. This is the so-called *space-time coincidence* K, which can be defined as follows: $K = R \cap L$. It is symmetric, reflexive, and transitive in S. The conical interpretation of K follows from the corresponding interpretations of the relations R and L.

The last absolute relation used here is the causal relation (physical interaction). It is sufficient here to use the so-called *unoriented causal relation* H'; $H'xy$ means that x is causally connected with y; hence the difference between cause and effect does not obtain here. The relation H' is symmetric and irreflexive in S.

With the help of the absolute time, space, and causal relations introduced above, I now define some important auxiliary concepts. These are necessary in order to construct eventistic definitions of the categories of things, processes, cross-sections, and coincidents.

Let **X** be an arbitrary set of events, i.e., $\mathbf{X} \subset \mathbf{S}$; then some concepts for it can be defined. First of all, the following two pairs of properties: (1) time extension (Ec) and time-non-extension ($\overline{Ec}$); (2) space extension (Ep) and space-non-extension ($\overline{Ep}$). They are defined as follows:

$$\mathbf{X} \in Ec \equiv \bigvee_{x,\, y \in \mathbf{X}} [\bar{R}xy],$$

$$\mathbf{X} \in \overline{Ec} \equiv \bigwedge_{x,\, y \in \mathbf{X}} [Rxy],$$

$$\mathbf{X} \in Ep \equiv \bigvee_{x,\, y \in \mathbf{X}} [Rxy \wedge \bar{L}xy],$$

$$\mathbf{X} \in \overline{Ep} = \bigwedge_{x,\, y \in \mathbf{X}} [Rxy \rightarrow Lxy].$$

That is, the set of events is time-extended iff at least two of its elements are time-separated, and the set of events is space-extended iff at least two of its quasi-simultaneous elements are space-separated.

From the viewpoint of the above-mentioned properties it is possible to divide the family of all sets of events, i.e., the power set $2^{\mathbf{S}}$, into four disjoint characteristic classes of sets of events. These classes are defined by the following products: (a) $Ec \cap Ep$, (b) $Ec \cap \overline{Ep}$, (c) $\overline{Ec} \cap Ep$, (d) $\overline{Ec} \cap \overline{Ep}$. This means that the set of events belonging to (a) is time- and space-extended, that belonging to (b) is time- but not space-extended, that belonging to (c) is space- but not time-extended, and that belonging to (d) is neither time- nor space-extended.

Secondly, for **X**-sets, the property of *time continuity Cn* can be defined, namely: $\mathbf{X} \in Cn$ iff **X** is ordered by the relation W and no section (in Dedekind's sense) of **X** is a jump or a gap. (*Nota bene*: The last three notions are defined by means of the relation W.)

Finally, I define for **X**-sets the property which I call *causal connectivity*, Cc:

$$\mathbf{X} \in Cc = \bigwedge_{x,\, y \in \mathbf{X}} [\bar{R}xy \rightarrow H'xy].$$

This means that the set of events has this property iff every two time-separated elements of it are causally connected.

This last property has nothing to do with Hume's conception of causality because—as we see later—only some special sets of events (certainly not **S**) have it. Of course, of all sets of events (including **S**) I assume the converse and universal statement: $\bigwedge_{x,\, y} [H'xy \rightarrow \bar{R}xy]$, which is called the *postulate of causality*.

All of the auxiliary concepts introduced above are absolute, i.e., independent of any arbitrary inertial reference system. This feature follows from the fact that the relations defining them are absolute.

By means of these concepts (properties), I can formulate eventistic definitions of the afore-mentioned categories of objects: things, processes, cross-sections, and coincidents.

The set of all things is denoted here by **T** and its elements by *a*, *b*, *c*, etc. Then the proposed eventistic definition of a thing is as follows:

(D1) $a \in \mathbf{T} \equiv a \subset \mathbf{S} \wedge a \neq \emptyset \wedge a \in (Ec \cap Ep) \wedge a \in Cn \wedge a \in Cc.$

In words: a thing is a non-empty set of events which is time- and space-extended, time-continuous, and causally connected.

The set of events which is defined only by means of the first three properties I call *thing-like* (**T**′). They are, of course, many sets of events which are thing-like, but are not things, e.g. the set **S**. Therefore we have the inclusion: $\mathbf{T} \subsetneqq \mathbf{T}'$.

Time-continuity, excluding jumps and gaps, is, in my view, a necessary condition of the so-called *identity of the thing through time* (*genidentity*). Hence, it is a very important property assumed for things. The same remark applies to the time-continuity of processes.

The application of this property to things presumes a linear (not only partial) ordering of them by the relation *W*; but this requires that things are treated as processes (as we see later). However, such schematization is in this respect admissible.

The set of all processes is denoted here by *Pr* and its elements by *f*, *g*, *h*, etc. The proposed eventistic definition of processes then is as follows:

(D2) $f \in Pr \equiv f \subset \mathbf{S} \wedge f \neq \emptyset \wedge f \in (Ec \cap \overline{Ep}) \wedge f \in Cn \wedge f \in Cc.$

In words: a process is a non-empty set of events which is time- but not space-extended, time-continuous, and causally connected.

The set of events which is defined by means of the first three properties only I call *process-like* (*Pr*′). There are many sets of events which are process-like, but are not processes. Hence we have the inclusion: $Pr \subsetneqq Pr'$.

The definition (D2) seems to be inadequate in the face of examples of many processes (in the everyday sense) which are space-extended (e.g. rivers). However, my reservation at this point is intentional and is enforced by an important theoretical aspect, namely, the need to draw a distinction between things and processes. The latter differ from things—according to my definitions—only in the property of space-non-extension. Therefore, rivers, being space-extended, are things.

The set of all cross-sections is denoted here by *Cr* and its elements by *q*, *r*, *s*, etc. The proposed eventistic definition is as follows:

(D3) $q \in Cr \equiv q \subset \mathbf{S} \wedge q \neq \emptyset \wedge q \in (\overline{Ec} \cap Ep)$.

In words: a cross-section is a non-empty set of events which is space- but not time-extended.

The sets of *Cr* can also—for symmetry—be called *cross-section-like* (*Cr′*); in this case, we have the equality: $Cr = Cr'$.

Cross-sections are commonly treated as time (instant) cross-sections of things. Hence the above definition might seem to be inadequate. However, the term as used here does not seriously conflict with current usage of the term 'cross-section', and therefore I can ignore this relativization of cross-section to things.

Finally, I denote the set of all *coincidents* by *Ks* and its elements by x', y', z', etc. The eventistic definition of these is as follows:

(D4) $x' \in Ks \equiv x' \subset \mathbf{S} \wedge x' \neq \emptyset \wedge x' \in (\overline{Ec} \cap \overline{Ep})$.

In words: a coincident is a non-empty set of events which is neither time- nor space-extended.

The sets of *Ks* may also—for the sake of symmetry—be called *coincident-like* (*Ks′*); in this case, we therefore have the equality: $Ks = Ks'$.

The concept of 'coincident' may seem somewhat artificial, especially as it does not appear in the language of physics. However, it is not merely a complement (from the point of view of time and space extensions) to concepts of things, processes, and cross-sections. As we see later, in the system of eventism some very important objects are in fact coincidents.

The four above-defined categories of objects: **T**, *Pr*, *Cr*, and *Ks*, obviously exclude one another. The corresponding classes (greater or equal): **T′**, *Pr′*, *Cr′*, and *Ks′*, cover the set $2^{\mathbf{S}}$ (minus $\emptyset$). These categories are absolute because all concepts which occur in their definitions are absolute, as was shown previously.

The properties *Ec*, *Ep* (and *Cn*, *Cc*) used above can be applied, by definition, only to the sets of events. Hence they cannot be applied to events themselves, which are individuals, i.e., non-sets. However, I said at the beginning that events are time- and space-extended, although in everyday, non-defined sense. Of course, we can stop at this point and say that we are dealing here with two different meanings of the analysed concepts. But we can also approach both by a convention: events are time- and space-non-extended, in the sense that their corresponding unit sets have just this property. In reality, the latter sets are coincidents which have the property $\overline{Ec} \cap \overline{Ep}$.

As I said earlier, the second type of relations of the set S, according to STR, are relative relations, i.e., relations dependent on an inertial reference system. Since these systems are certain special things, the relations are relativized to certain things. Hence, we must already have available here the concept of 'thing' and, as we know, we have it.

The relative time relations are: (1) *earlier than* (earlier) W_u and *later than* (later) $\breve{W}_u$; both are asymmetric and transitive in S; (2) *simultaneity* R_u, where $R_u = \overline{W_u} \cap \overline{\breve{W}_u}$; it is symmetric, reflexive, and transitive in S, and is therefore an equivalence in S. The index u represents one a certain (definite) reference system.

The relative space relations are: (1) *space separation* $\overline{L_u}$, which is symmetric and irreflexive in S; (2) *co-location* L_u; it is also an equivalence relation in S—as is the relation R_u.

The *space-time coincidence* relation K', where $K' = R_u \cap L_u$, is not relative but absolute (hence it is without index u). Therefore, it is equal to coincidence K, i.e., $K' = K$; it is also an equivalence relation in S.

On the basis of the relative relations introduced above, it is easy to construct eventistic definitions of some very important kinds of objects such as moments, space points, and space-time points (or points, for short). This is done by means of the same procedure, namely definition by abstraction, because R_u, L_u, and K' are all equivalence relation in the set of all events S.

Let C_u, P_u, and CP be sets of moments, space points, and points, respectively, represented by m, p', and p. Then I define the last as follows (x is an arbitrary event):

(D5) $m \in C_u \equiv \bigvee_x [m = |x|_{R_u}]$,

(D6) $p' \in P_u \equiv \bigvee_x [p' = |x|_{L_u}]$,

(D7) $p \in CP \equiv \bigvee_x [p = |x|_{K'}]$.

In words: moments are abstraction classes of simultaneity R_u in S, space points are abstraction classes of co-location L_u in S, and points are abstraction classes of space-time coincidence K' in S.

As such classes, they are special non-empty sets of events, i.e., parts of set S. According to the nature of the relations R_u, L_u, and K', moments and space points are relative (to the system u), while points, however, are absolute.

If we assume that events—as individuals—have the logical type 0, then things, processes, cross-sections, and coincidents, on the one hand, and moments, space points and points, on the other, as sets of events, are of the same logical type 1. One can ask if it is possible to point out certain important sets founded in events which have a higher logical type than 1. Of course, it is.

To begin with, if we do not go deeper into the structure of time, space, and space–time, then we can define time as the set of all moments, i.e., C_u, space as the set of all space points, i.e., P_u, and space–time as the set of all points, i.e., CP. All these objects are families of some special corresponding sets of events. As such, they are all of logical type 2. Secondly, if I assume (generally) that properties and relations are sets, and that laws (e.g. physical) are relations between certain properties of things, then laws are eventistic sets of logical type 4; meta-laws (as properties of laws) are also eventistic sets and of logical type 5.

An important question is: how (in STR) are absolute and relative time and space relation interconnected? The answer is well known and simple (**U** = the set of all inertial reference systems):

(T1) $W = \bigcap_{u \in \mathbf{U}} W_u.$

In words: the absolute relation W is the intersection of all definite relative relations W_u: $W_{u_1}, W_{u_2}, \ldots, W_{u_n}$.

(T2) $R = \bigcup_{u \in \mathbf{U}} R_u.$

In words: the absolute relation R is the union of all definite relative relations R_u: $R_{u_1}, R_{u_2}, \ldots, R_{u_n}$. Analogous connections hold between relations $\bar{L}$ and $\bar{L}_u$ and L and L_u, respectively.

The treatment of these statements as definitions of absolute relations by means of the corresponding relative relations is incorrect, because then we fall into a *circulus in definiendo*. The set of system **U** is a subset of things **T** which are defined in eventism with the help of absolute relations. Therefore, we must treat these statements here as axioms or theses of STR.

Independently of the latter, the connections (T1) and (T2) allow us to establish a relationship between processes, cross-sections, and coincidents, on the one hand, and moments, space points, and points, on the other. First, moments are cross-sections, i.e., $C_u \subsetneq Cr$. In fact, due to (D5), every two events of a moment m are simultaneous, therefore they are quasi-simultaneous, since $R_u \subset R$ (T2), ergo m is time-non-extended (but space-extended). Secondly, space points are certain process-like objects, i.e., $P_u \subsetneq Pr'$. Due to (D6), every two events of space point p'

are co-local, therefore they are quasi-co-local, since $L_u \subset L$ (analogue of (T2)), ergo p' is space-non-extended (but time-extended). Third, points are certain coincidents, i.e., $CP \subsetneq Ks$. Due to (D7), every two events of a point p enter into space-time coincidence K', ergo into space-time coincidence K; hence p is space- and time-non-extended.

In conclusion, I should like to emphasize that point-eventism as an ontology differs from the eventism of B. Russel (Russel, 1948) and A. N. Whitehead (Whitehead, 1919), who regard events as objects that are time- and space-extended (although only slightly).

In this connection, I shall prove an important logical implication between two statements: (1) the simultaneity R_u is equivalence in **S** and therefore a transitive relation; and (2) events are instantaneous, i.e., are time-non-extended. Suppose, in contradiction to (2), that our events are time-extended and that R_u means a total or partial time-overlapping of them. Then it is always possible to find three events x, y, z, such that we have $R_u(xy) \wedge R_u(yx) \wedge \sim R_u(xz)$; in this case, the simultaneity R_u is not transitive, i.e., we have the negation of (1). Therefore, the transitivity of simultaneity (1) entails the time punctuality (time-non-extension) of events (2). Exactly the same holds for the relation of co-location L_u; its transitivity entails the space punctuality of events if we accept that L_u means a total or partial space-overlapping of events, of course.

In physics, especially in STR (but also in classical mechanics), relations R_u and L_u are treated as equivalence, hence transitive relations in the set **S** of all events. Hence the assumption that events are time- and space-non-extended is one of the basic assumptions of physics, not the idle speculation of some philosophers.

The answer to the question of why physics makes this assumption is easy to give: without it, it is impossible to define moments and space points (at least in the above-mentioned natural way), which are necessary for the measurement of time and space intervals.

BIBLIOGRAPHY

Augustynek, Z. (1968), Homogeneity of time, *American Journal of Physics*, **36**, 126–132, (1970), *Properties of Time*, Warsaw: PWN (in Polish); (1975), *The Nature of Time* Warsaw: PWN (in Polish); (1976), Past, present, and future in relativity, *Studia Logica* XXXV, 45–53; (1976a), Relational becoming *Poznań Studies in the Philosophy of Sciences and the Humanities* **2** (2), 12–23; (1979), *Past, Present, Future*. Warsaw: PWN (in Polish); (1982), Time Separation, in W. Krajewski (ed.) *Polish Essays in the Philosophy of the Natural Sciences*, Dordrecht: Reidel, pp. 215–222; (1987), Point-eventism, *Reports on Philosophy*, **11**, 49–55.

Borsuk, K. and Szmielew, W. (1972), *Foundation of Geometry*. Warsaw: PWN (in Polish).

Broad, C. D. (1923), *Scientific Thought*. London: Kegan Paul.

Carnap, R. (1929), *Abriss der Logistik*. Vienna: Springer.

Christenson, J., Cronin, J. W., Fitch, V. L., and Turlay, R. (1964), Evidence for 2π decay of the K_2 meson. *Phys. Rev. Lett.* **13**, 138.

Gale, R. (1968), *The Language of Time*. New York: Humanities Press.

Goodman, N. (1951), *The Structure of Appearance*. Cambridge: Harvard Univ. Press.

Kotarbiński, T. (1913), *Practical Essays*. Warsaw: Kasa im. Mianowskiego (in Polish).

Łukasiewicz, J. (1961), *On Problems of Logic and Philosophy. Selected Papers*. Warsaw: PWN (in Polish).

McTaggart, J. M. E. (1927), *The Nature of Existence*. Vol. II, Cambridge: Cambridge University Press.

Mehlberg, H. (1980), *Time, Causality, and the Quantum Theory*. Dordrecht: Reidel.

Mostowski, A. (1948), *Mathematical Logic*. Warszawa–Wrocław (in Polish).

Putnam, H. (1967), Time and physical geometry. *Journal of Philosophy* **64**, 240–247.

Quine, W. V. O. (1953), Mr. Strawson on logical theory. *Mind LXII*, 433–451.

Reichenbach, H. (1924), *Axiomatic der relativistischen Raum-Zeit-Lehre*. Braunschweig: Vieweg and Sons; (1928), *Philosophie der Raum-Zeit-Lehre*. Berlin: Walter de Gruyter and Company.

Russell, B. (1903), *The Principles of Mathematics*. Cambridge: Cambridge University Press; (1948), *Human Knowledge*. London: Allen and Unwin.

Sellars, W. S. (1962), Time and the world order. *Minnesota Studies in the Philosophy of Science*, Vol. III. Minneapolis: University of Minnesota Press.

Strawson, P. F. (1952), *Introduction to Logical Theory*. London: Methuen.

Whitehead, A. N. (1919), *Enquiry Concerning the Principles of Natural Knowledge*. Cambridge: Cambridge University Press.

INDEX

Nijhoff International Philosophy Series

1. N. Rotenstreich: *Philosophy, History and Politics.* Studies in Contemporary English Philosophy of History. 1976 ISBN Pb 90-247-1743-4
2. J. T. J. Srzednicki: *Elements of Social and Political Philosophy.* 1976 ISBN Pb 90-247-1744-2
3. W. Tatarkiewicz: *Analysis of Happiness.* Translated from Polish by E. Rothert and D. Zieliński. 1976 ISBN 90-247-1807-4
4. K. Twardowski: *On the Content and Object of Presentations.* A Psychological Investigation. Translated from Polish with an Introduction by R. Grossmann. 1977 ISBN Pb 90-247-1926-7
5. W. Tatarkiewicz: *A History of Six Ideas.* An Essay in Aesthetics. Translated from Polish a.o. by C. Kasparek. 1980 ISBN 90-247-2233-0
6. H. W. Noonan: *Objects and Identity.* An Examination of the Relative Identity Thesis and Its Consequences. 1980 ISBN 90-247-2292-6
7. L. Crocker: *Positive Liberty.* An Essay in Normative Political Philosophy. 1980 ISBN 90-247-2291-8
8. F. Brentano: *The Theory of Categories.* Translated from German by R. M. Chisholm and N. Guterman. 1981 ISBN 90-247-2302-7
9. W. Marciszewski (ed.): *Dictionary of Logic* as Applied in the Study of Language. Concepts - Methods - Theories. 1981 ISBN 90-247-2123-7
10. I. Ruzsa: *Modal Logic with Descriptions.* 1981 ISBN 90-247-2473-2
11. P. Hoffman: *The Anatomy of Idealism.* Passivity and Activity in Kant, Hegel and Marx. 1982 ISBN 90-247-2708-1
12. M. S. Gram: *Direct Realism.* A Study of Perception. 1983 ISBN 90-247-2870-3
13. J. T. J. Srzednicki, V. F. Rickey and J. Czelakowski (eds.): *Leśniewski's Systems.* Ontology and Mereology. 1984 ISBN 90-247-2879-7
    For 'The Leśniewski Collection' *see also Volume 24.*
14. J. W. Smith: *Reductionism and Cultural Being.* A Philosophical Critique of Sociobiological Reductionism and Physicalist Scientific Unificationism. 1984 ISBN 90-247-2884-3
15. C. Zumbach: *The Transcendent Science.* Kant's Conception of Biological Methodology. 1984 ISBN 90-247-2904-1

Volumes 1–8 previously published under the Series Title: Melbourne International Philosophy Series.

Nijhoff International Philosophy Series

16. M. A. Notturno: *Objectivity, Rationality and the Third Realm: Justification and the Grounds of Psychologism.* A Study of Frege and Popper. 1985 ISBN 90-247-2956-4
17. I. Dilman (ed.): *Philosophy and Life.* Essays on John Wisdom. 1984 ISBN 90-247-2996-3
18. J. J. Russell: *Analysis and Dialectic.* Studies in the Logic of Foundation Problems. 1984 ISBN 90-247-2990-4
19. G. Currie and A. Musgrave (eds.): *Popper and the Human Sciences.* 1985 ISBN Hb 90-247-2998-X; Pb 90-247-3141-0
20. C. D. Broad: *Ethics.* Lectures given at Cambridge during the Period 1933–34 to 1952–53. Edited by C. Lewy. 1985 ISBN 90-247-3088-0
21. D. A. J. Seargent: *Plurality and Continuity.* An Essay in G. F. Stout's Theory of Universals. 1985 ISBN 90-247-3185-2
22. J. E. Atwell: *Ends and Principles in Kant's Moral Thought.* 1986 ISBN 90-247-3167-4
23. J. Agassi, and I. C. Jarvie (eds.): *Rationality: The Critical View.* 1987 ISBN Hb 90-247-3275-1; Pb 90-247-3455-X
24. J. T. J. Srzednicki and Z. Stachniak: *S. Leśniewski's Lecture Notes in Logic.* 1988 ISBN 90-247-3416-9
    For 'The Leśniewski Collection' *see also Volume 13.*
25. B. M. Taylor (ed.): *Michael Dummett.* Contributions to Philosophy. 1987 ISBN 90-247-3463-0
26. A. Z. Bar-On: *The Categories and the Principle of Coherence.* Whitehead's Theory of Categories in Historical Perspective. 1987 ISBN 90-247-3478-9
27. B. Dziemidok and P. McCormick (eds.): *On the Aesthetics of Roman Ingarden.* Interpretations and Assessments. 1989 ISBN 0-7923-0071-8
28. J. T. J. Srzednicki (ed.): *Stephan Körner — Philosophical Analysis and Reconstruction.* Contributions to Philosophy. 1987 ISBN 90-247-3543-2
29. F. Brentano: *On the Existence of God.* Lectures given at the Universities of Würzburg and Vienna (1868–1891). Edited and translated from German by Susan F. Krantz. 1987 ISBN 90-247-3538-6
30. Z. Augustynek: *Time: Past, Present, Future.* Essays dedicated to Henryk Mehlberg. 1990 ISBN 0-7923-0270-2
31. T. Pawlowski: *Aesthetic Values.* 1989 ISBN 0-7923-0418-7

Nijhoff International Philosophy Series

32. M. Ruse (ed.): *What the Philosophy of Biology Is.* Essays dedicated to Davis Hull. 1989 ISBN 90-247-3778-8
33. J. Young: *Willing and Unwilling.* A study in the Philosophy of Arthur Schopenhauer. 1987 ISBN 90-247-3556-4
34. T. Z. Lavine and V. Tejera (eds.): *History and Anti-History in Philosophy.* 1989 ISBN 0-7923-0455-1
35. R. L. Epstein: *The Semantic Foundations of Logic.* Volume 1: Propositional Logics. With the assistance and collaboration of W. A. Carnielli, I. M. L. D'Ottaviano, S. Krajewski and R. D. Maddux. 1990 ISBN 0-7923-0622-8
36. A. Pavković (ed.): *Contemporary Yugoslav Philosophy.* The Analytic Approach. 1988 ISBN 90-247-3776-1
37. A. Winterbourne: *The Ideal and the Real.* An Outline of Kant's Theory of Space, Time and Mathematical Construction. 1988 ISBN 90-247-3774-5
38. K. Szaniawski (ed.): *The Vienna Circle and the Lvov-Warsaw School.* 1989 ISBN 90-247-3798-2
39. G. Priest: *In Contradiction.* A Study of the Transconsistent. 1987 ISBN 90-247-3630-7

Further information about our publications on *Philosophy* are available on request.

Kluwer Academic Publishers—Dordrecht /Boston/London

MIX
Papier aus verantwortungsvollen Quellen
Paper from responsible sources
FSC® C105338

If you have any concerns about our products,
you can contact us on
ProductSafety@springernature.com

In case Publisher is established outside the EU,
the EU authorized representative is:
Springer Nature Customer Service Center GmbH
Europaplatz 3, 69115 Heidelberg, Germany

Printed by Libri Plureos GmbH
in Hamburg, Germany